RESEARCH ON ENERGY ISSUES IN CHINA

RESEARCH ON ENERGY ISSUES IN CHINA

JIANG ZEMIN

Professor, Shanghai Jiao Tong University
Former President, People's Republic of China

Translated by: The Central Translation Bureau, Beijing

Translators: Jia Yuling, Alan A. Johnston, Liu Liang, Richard A. O'Connell, Sun Xianhui, Tong Dongjie, Tong Xiaohua, Wang Lili, and Zhu Yanhui

ELSEVIER

AMSTERDAM • BOSTON • HEIDELBERG • LONDON • NEW YORK • OXFORD
PARIS • SAN DIEGO • SAN FRANCISCO • SINGAPORE • SYDNEY • TOKYO
Academic Press is an imprint of Elsevier

Editorial Director, Asia Pacific, Elsevier Science & Technology Books: *Denise E.M. Penrose*
Vice President, Elsevier Science & Technology Books, Asia-Pacific: *Sanjiv Pandya*
Managing Director, Elsevier Science & Technology China: *Zhang Yuguo*
Managing Director, Elsevier Science & Technology Books, China: *Ying Zhongfeng*

Academic Press is an imprint of Elsevier.
30 Corporate Drive, Suite 400, Burlington, MA 01803, USA
525 B Street, Suite 1900, San Diego, CA 92101-4495, USA
Radarweg 29, PO Box 211, 1000 AE Amsterdam, The Netherlands
Linacre House, Jordan Hill, Oxford OX2 8DP, UK
32 Jamestown Road, London NW1 7BY, UK

First published in Chinese.
Chinese edition, copyright © 2008 Shanghai Jiao Tong University Press

江泽民著《中国能源问题研究》中文版由上海交通大学出版社于 2008 年 10 月出版。

The Chinese version of *Research on Energy Issues in China* by Jiang Zemin is published by Shanghai Jiao Tong University Press in Oct., 2008.
English translation copyright © 2010 Elsevier Inc. All rights reserved.
This English translation of *Research on Energy Issues in China* by Jiang Zemin is published by arrangement with Shanghai Jiao Tong University Press, which undertook the translation.

Library of Congress Cataloging-in-Publication Data
A catalog record for this book is available from the Library of Congress

British Library Cataloguing in Publication Data
A catalogue record for this book is available from the British Library

Chinese edition ISBN: 978-7-313-05301-5
English translation ISBN: 978-0-12-378619-7

For information on all Academic Press publications
visit our website at elsevierdirect.com

Transferred to Digital Printing in 2013

Working together to grow
libraries in developing countries

www.elsevier.com | www.bookaid.org | www.sabre.org

ELSEVIER   BOOK AID
            International   Sabre Foundation

CONTENTS

Speech at a Meeting with Experts Attending a Forum on "Reflections on Energy Issues in China"

Jiang Zemin with experts attending a forum on "Reflections on Energy Issues in China" on April 9, 2008.

Most of you participating in today's forum are experts on energy issues, including a number of my former classmates and colleagues, as well as the Party secretary and president of my alma mater, Shanghai Jiao Tong University. This forum presents a rare opportunity for you to carry out academic exchanges on a topic of common interest, and I wish to express my sincere thanks for your attendance. We have just heard reports from Comrade Ma Dexiu; Li Jinghai, vice president of the Chinese Academy of Sciences; Du Xiangwan, vice president of the Chinese Academy of Engineering; and Comrade Zhang Guobao on how the academic world views my article, "Reflections on Energy Issues in China." It now seems that most of the opinions expressed in it have been met with general acceptance, although of course there may also be some differing views.

It took nearly six months to write this article, during which time I invited a number of experts to discuss it with me on fifteen separate occasions; Ma Fucai, Ning Jizhe, Zhou Dadi, Han Wenke, and other comrades also helped with specific work on it, and I accordingly convey my thanks to all of them.

I. ORIGINS OF THE ARTICLE

Why did I write this article? Its origins can be traced back to when I was studying in the Soviet Union. My tutor, M. I. Trehov, published a book on energy conservation at machinery manufacturing plants, and after returning to China, I translated it into Chinese in my spare time. However, as a result of the Cultural Revolution and other reasons, my translation was not published at that time.

While serving as director of the power plant at the First Automotive Works (FAW) in Changchun, my work directly concerned energy issues. This was a time when I integrated theory with practice and began to study and think about energy and energy conservation. Later, while working in Shanghai, I was appointed a professor at Shanghai Jiao Tong University and published "Trends in Energy Development and Important Energy Conservation Measures" in the *Journal of Shanghai Jiao Tong University* in 1989. This article chiefly reviewed the history of worldwide energy development, forecast its future course, and drew on my experience with FAW to set forth a number of energy conservation measures.

During the last 20 years, the field of energy resources has undergone tremendous change, but the goal of energy conservation has not, and the basic nature of energy conservation in organizations remains the same. Therefore, some of the thinking and methods I advanced in that article are still worth revisiting. Last year, when the *Journal of Shanghai Jiao Tong University* solicited another article from me, I thought about the importance of energy issues and decided to update the article I had published by adding a survey of my views on energy issues that have emerged since then. This resulted in the article "Reflections on Energy Issues in China," published in the *Journal of Shanghai Jiao Tong University* before the 2008 anniversary of the university's founding.

II. THE GREAT IMPORTANCE OF ENERGY ISSUES

Energy is the food of industry and the lifeblood of the national economy. Energy issues not only represent major economic and social problems; they also involve critical diplomatic, environmental, and security issues. In the 1970s, oil crises twice led to global economic recessions. Since the beginning of this century, the continuing climb in oil prices has had a significant effect on the global economy, especially on oil-importing countries. Today, energy security is increasingly a global issue, and all major countries now view solving energy issues as an important aspect of their national strategy and common policy.

China is in a period of accelerating industrialization and urbanization, and its energy demands are constantly growing. However, the country's per capita energy resources are fewer than the world average; its energy efficiency is not high; the main energy resource China consumes is coal; there is great pressure on the ecosystem; and the energy resource structure still requires further optimization. Together, these facts dictate that the demand for energy will continue to outstrip supply and that contradictions in the structure of energy resources will persist for a considerable time to come. Given that the assurance of a steady supply of energy is a strategic issue confronting our country's development, we must accord energy an important place in our country's overall development strategy and focus on it unequivocally.

III. FOCUSING ON RESEARCH INTO ENERGY ISSUES

It is possible that within the next 10 to 20 years China will become the world's largest energy consumption and supply system; therefore, we must seriously consider which energy strategy to implement and which energy technology to choose. In this article I argue that we must pursue a new, distinctively Chinese, energy development path, and I discuss developing and using coal, natural gas, and oil, as well as new and renewable energy sources such as hydro, thermal, nuclear, wind, and solar power. The general line of thought is that we need to be steadfast in our conservation of energy, use it efficiently, diversify development, keep the environment clean, be technology-driven, and pursue international cooperation in order to establish a system of energy production, distribution, and consumption that is highly efficient, uses advanced technology, produces few pollutants, has minimal impact on the ecosystem, and provides a steady and secure energy supply.

Of course, these explorations are still in the preliminary stages, and many problems still need thorough investigation. For example, to optimize the structure of energy sources, we must further research the clean use of coal, press ahead with the development of new and renewable energy sources, and optimize the electricity supply system. To make scientific progress in the field of energy, we must further research, develop, and utilize advanced technology, for example by prospecting for new oil and gas resources, using energy cleanly and efficiently, and developing advanced new energy sources for the future. To conserve energy, we must further research how to advance conservation technology, improve energy conservation management, and raise energy efficiency across the board. To ensure energy security, we must research how to make good use of domestic and foreign energy resources, implement energy diversification, strengthen energy forecasting and early warnings, and increase our ability to respond to emergency situations to ensure a steady supply of energy.

IV. STRESSING RESEARCH INTO ENERGY POLICIES

Sound energy policy is an important requirement for the implementation of an energy strategy. Conversely, the lack of good policies makes it very difficult to implement such a strategy, no matter how good it is. In the article I advance a number of suggestions: widening the application of new energy technology and products, stressing resource conservation and environmental protection, actively adopting measures to address climate change, making major energy facilities more future-oriented, increasing funding through a variety of channels for energy exploitation and technology innovation, creating a reasonable mechanism for setting prices for energy resources, and improving the system and mechanisms for energy management. Of course, energy policies cannot be improved overnight. The task requires us to proceed based on an understanding of China's realities, draw on foreign experience, emancipate our minds, keep up with the times, and constantly explore new thinking, mechanisms, and methods for developing and managing energy.

All of you gathered here are experts in the energy field, and I hope you continue to study energy issues deeply and generate valuable research results. Academics, the government, and enterprises need to interact continually to cooperate with one another, pool their collective wisdom, join their efforts, find practical applications for more research results, and thereby promote the sound development of our country's energy cause.

Reflections on Energy Issues in China*

Jiang Zemin addressing experts attending a forum on "Reflections on Energy Issues in China" on April 9, 2008.

*Originally published in the *Journal of Shanghai Jiao Tong University*, No. 3, 2008.

Energy is a vital material foundation for the existence and development of humankind and is currently a focal point in international politics, economics, military affairs, and diplomacy. Energy security is indispensable to China's sustained and rapid economic and social development. As economic globalization increases and the pace of China's modernization accelerates, the way in which we understand energy development trends, which energy development strategy we select, and which policies and measures we adopt are all extremely important issues requiring serious consideration.

I. SIGNIFICANCE OF ENERGY ISSUES

Human use of energy has evolved from the firewood era to the coal era, and on to the oil and gas era. As total energy consumption has continued to grow, so too the energy structure has continually changed (see Figures 1 and 2). The onset of each energy era has brought with it huge increases in the productive forces, greatly boosting the world's economic and social development. At the same time, growth in energy consumption, especially fossil fuel consumption, constraints of the energy supply on economic and social development, and the impact of energy consumption on the environment have become more pronounced.

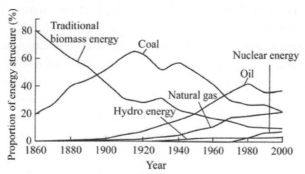

Figure 1   Changes in the world's energy resource structure since 1860[1]

---

[1]Gilbert Jenkins, *Oil Economists' Handbook*, 5th edition, Elsevier Applied Science, London, 1989, and World Energy Council, *Global Energy Perspectives to 2050 and Beyond*, World Energy Council, London, 1995.

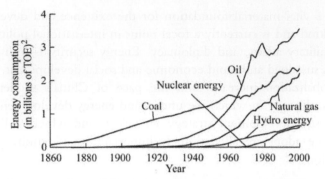

Figure 2 Changes in the world's energy consumption since 1860 [1]

From the perspective of modern economic and social development, the importance of energy issues primarily manifests itself in the following four ways.

1. Energy as a material base for modern economic and social development

Modern economic and social development is based on a high level of material, cultural, and ethical progress. Achieving this high level of material progress requires a huge expansion of the productive forces, as well as modern agricultural, industrial, transportation, and logistics systems and living and service facilities—all of which depend on energy. In modern society the proportion of energy consumed in food production for sustaining life has significantly dropped, whereas industrial production, daily life, and transportation services have become major areas of energy consumption. The development course of developed countries indicates that the energy consumption of a country during the early and middle stages of industrialization typically experiences a period of rapid growth, with an energy consumption elasticity coefficient[2] greater than 1. During the late

---

[2]The energy consumption elasticity coefficient reflects the relationship between economic growth and an increase in energy consumption. It is defined as the ratio between the increased rate of energy consumption and the growth rate of the national economy. The formula is as follows: energy consumption elasticity coefficient equals average annual growth rate of energy consumption divided by average annual growth rate of the national economy.

stage of industrialization, or the post-industrial stage, energy consumption generally undergoes a period of slow growth, with an energy consumption elasticity coefficient less than 1. History further demonstrates that when the per capita GDP of a country or region reaches a certain level, the energy consumed for food, clothing, housing, transportation, and daily necessities will rise, and on average people will consume significantly more energy in their daily lives. It is fair to say that without an adequate energy supply to sustain it, no modern society or civilization would be possible.

2. Energy as a major constraint on economic and social development

Since the 1950s, China's energy industry has steadily grown and developed. In particular, since adoption of the reform and opening up policy, the country's energy supply capacity has constantly grown, which in turn contributed to sustained and rapid economic growth. However, in the course of this economic development, energy shortages became an acute problem. Generally, once the scale of investment in fixed assets expanded and economic development accelerated, shortages emerged in the supply of coal, electricity, oil, and transportation, resulting in a bottleneck restricting economic and social development. Toward the end of the 1990s, due to progress in the market-oriented reform of the energy sector, further opening up of the energy industry to the outside world, and rising investment, coal and electricity output increased enormously—as did oil and natural gas imports—thus greatly alleviating the constraints the energy supply had imposed on economic and social development. Since the beginning of the twenty-first century, however, further changes have occurred in energy supply and demand. The acceleration of industrialization and urbanization, coupled with the overdevelopment of energy-intensive industries, has driven energy demand to historic highs (see Figure 3), causing the constraints energy imposes on economic and social development to increase again. China, as a developing country with a huge population, is still a long way from attaining a relatively high level of modernization. As the economy and society continue to develop and people's living standards constantly improve, energy demand will continue to grow, and constraints arising from the imbalance between energy supply and demand and energy-related environmental issues will persist.

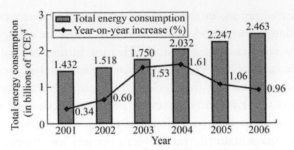

Figure 3   Increase in China's energy consumption since 2001 [2]

3. Energy security as a vital element in economic and national security

Oil security is the most important factor in energy security. The two global oil crises in the 1970s led to economic slowdowns in the leading developed countries and fluctuations in the global economy. Since the beginning of the twenty-first century, oil prices have continued to climb (see Figure 4), and at the beginning of 2008, the price of crude oil futures exceeded $100 per barrel. The increase in oil prices has exerted a considerable influence on the global economy, especially on the economies of countries heavily dependent on oil imports. In some countries inflated oil prices have even triggered social unrest. As the history of industrialization shows, developed countries used great quantities of foreign energy resources in addition to domestic resources in the course of their industrialization. Even today, many developed countries still rely heavily on

Figure 4   The upward trend of oil prices in the international market[3]

---

[3]Energy Information Administration, U.S. Department of Energy.

imported oil and natural gas. In today's increasingly globalized economy, the allocation of energy resources on a global scale is the prevailing trend. However, the inequitable international political and economic order and energy market regulations create a number of barriers to the use of international resources by developing countries, and energy issues are an important factor in a number of regional conflicts and local wars. Given its limited domestic energy resources, China has to make use of both international and domestic markets to secure its energy supplies, especially in oil and natural gas. Currently, China depends on external sources for nearly 50 percent of its oil supply, and this proportion is expected to rise. Therefore, the international oil market will exert an ever greater influence on China's energy, economic, and even national security.

4. The increasingly prominent impact of energy consumption on the ecosystem

While promoting world development, the exploitation and utilization of energy resources have also brought about serious ecological and environmental problems. The use of fossil fuels is a major cause of increases in CO_2 and other greenhouse gases. Scientific observations indicate that the concentration of CO_2 in the atmosphere has increased from 280 ppmv (1 ppmv = 10^{-6}) prior to the Industrial Revolution to 379 ppmv at present (see Figure 5), and that the average world temperature has risen by

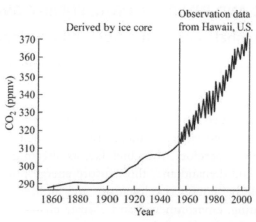

Figure 5   Changes in the atmospheric concentration of CO_2 since 1860 [3]

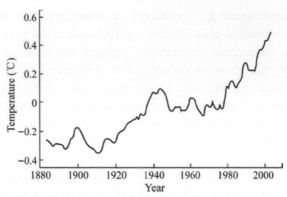

Figure 6 Changes in average global temperature since 1880 [4]

0.74°C over the past century (see Figure 6), with a noticeable increase in the past 30 years in particular. The impact of global warming on natural ecosystems and the environment that humans rely on for their existence is generally negative, and this trend requires serious action on the part of the international community.

In China coal has long been the main energy resource. However, SO_2, CO_2, and particulate matter emitted in the course of the production and use of coal are a major source of air pollution and greenhouse gases. Therefore, the solution to China's energy issues is about more than just striking a proper balance between supply and demand; it also involves addressing the ecological and environmental concerns raised by large-scale energy use.

II. DOMESTIC AND INTERNATIONAL ENERGY OVERVIEW

1. The basic global energy situation and development trends

In recent years, with global energy consumption and oil prices constantly rising, people have become increasingly worried about the sustainability of global energy supplies. At present the world's energy supply is heavily dependent on fossil fuels. Its recoverable fossil fuel reserves will last for a relatively long time, and currently there are no substantive constraints on the energy supply. Therefore, the chief factors affecting the relationship between supply and demand and, thus, future energy prices in the global marketplace will be energy exploitation and utilization technologies, energy restructuring, environmental and climate change, and the international political and economic order.

a. The world has abundant fossil energy reserves, but they are distributed unequally among countries

At the end of 2006 the world's proven recoverable coal reserves totaled 909.1 billion tons,[4] enough for 147 more years of exploitation at current production levels. Although global reserves of conventional oil and natural gas are lower than those of coal, new deposits are discovered every year. Over the past 20 years the reserves-to-production ratios of oil and natural gas have remained between 40 and 60, with little variation (see Figure 7). Moreover, the world has abundant unconventional oil and natural gas resources that are presently unsuited to large-scale development, either because they cannot be exploited using existing technology or because the cost of doing so is prohibitive. Examples of such resources include heavy oil (crude oil with a density of $0.920-1.000$ g/cm^3), oil sands,[5] shale oil,[6] and

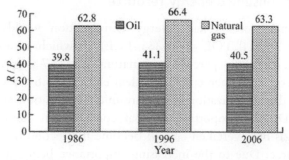

Figure 7 Changes in the reserves-to-production (R/P) ratios of the world's oil and natural gas resources[7]

---

[4]*BP (British Petroleum) Statistical Review of World Energy*, various editions 1979–2007. This periodical publishes relatively comprehensive statistics on world energy commodities.

[5]Oil sands are sands from which oil can be extracted; they are a mixture of sand, water, and asphalt, with the percentage of asphalt typically ranging from 6 percent to 20 percent. Oil-sand asphalt is an organic substance composed of both hydrocarbon and nonhydrocarbon materials.

[6]Shale oil is crude oil distilled from shale. Shale is a kind of high-ash combustible organic rock, composed of multiple inorganic and organic minerals. It is similar in appearance to mud shale and typically has an oil content of 4 percent to 20 percent, with a maximum of 30 percent. When heated to 500°C, the organic substances in the shale will decompose into shale oil, which has characteristics close to crude oil.

[7]BP world energy statistics.

gas hydrates,[8] which have immense potential for future exploitation and utilization. However, known energy resources are distributed very unevenly throughout the world. Coal resources are concentrated mainly in the United States, Russia, China, India, and Australia. Although oil resources can be found on every continent, they are concentrated largely in the Middle East and a few other countries. The proven recoverable oil reserves of the Organization of Petroleum Exporting Countries (OPEC) make up 75.7 percent of the world's total, with reserves of Middle Eastern countries accounting for more than 60 percent. In terms of individual countries, the top 10 countries with recoverable oil reserves possess 82.6 percent of the world's total. Similarly, natural gas resources are concentrated mainly in the Middle East, Russia, and Central Asia, with Russia, Iran and Qatar possessing 55.7 percent of the world's total reserves.

b. The energy structure is diversifying, but fossil fuels remain the most consumed energy resource

In 2006 primary commercial energy consumption worldwide totaled 10.88 billion TOE (1 TOE = 1.4286 TCE), of which oil accounted for 35.8 percent, coal 28.4 percent, and natural gas 23.7 percent, followed by hydro energy at 6.3 percent and nuclear energy at 5.8 percent.[9] Among members of the Organization for Economic Cooperation and Development (OECD), the proportion of coal consumption continues to fall, and natural gas has overtaken coal to become the second most consumed energy resource. Due to the increasing importance being given to environmental issues by the international community, as well as the continuing development of energy technologies, clean forms of energy to replace coal and oil are maturing rapidly. These new technologies will further reduce the proportion of coal and oil in total primary energy demand and increase that of natural gas, nuclear energy, and renewable forms of energy. However, given the technological and economic factors currently restraining the

---

[8]Gas hydrate is an unconventional natural gas resource that has drawn much attention in recent years. It is a crystal hydrate of water and methane that resembles ice and has strong adsorption capacity due to its supramolecular structure. Distributed mainly in continental shelf areas under certain high-pressure and low-temperature conditions, it is also found in the permafrost region of some continents. Gas hydrate is highly flammable because it has high methane or other hydrocarbon content, so it is also called "combustible ice."

[9]*BP Statistical Review of World Energy,* various editions 1979–2007.

development of nuclear, wind, solar, and biomass energy, it will be a long time before nonfossil-based types of energy replace fossil-based energy on a large scale. It is estimated that fossil fuels such as oil, natural gas, and coal will remain the primary energy resources up until 2030.

c. Energy consumption in developed countries remains high, and energy demand in developing countries is growing rapidly

In the course of their industrialization and post-industrialization, developed countries created industrial, transportation, and building systems that consume high quantities of energy. In 2006 OECD countries accounted for 51 percent of the world's total energy consumption, with per capita consumption of 4.74 TOE. The United States had the highest per capita consumption, 7.84 TOE, while the figure for China was 1.31 TOE; for African countries it was a mere 0.36 TOE (see Figure 8). Developed countries are already in a period of relatively slow growth in energy consumption, whereas the energy consumption of developing countries, which are still accelerating growth to end poverty and backwardness, is increasing at an ever faster rate. Between 1996 and 2006 the average annual increase in energy consumption in 26 European and North American countries was 0.62 percent, compared with 4.36 percent in developing countries. The International Energy Agency (IEA) predicts that total world energy demand will grow at an average annual rate of 1.2 percent to 1.6 percent from 2006 to 2030, with 70 percent of that growth coming from developing countries (see Figure 9).

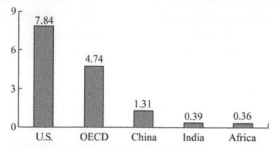

Figure 8   Per capita energy consumption of several countries and regions in 2006 (TOE/person)[10]

---

[10]BP (2007) and IEA (2007).

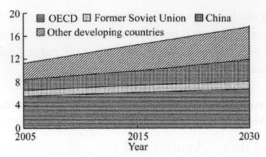

Figure 9 Forecasts for energy demand in different regions (in billions of TOE)[11]

d. The impact of climate change on energy development is increasing, and low-carbon and carbon-neutral sources of energy have become the new focus of attention

As people become increasingly aware of the relationship between greenhouse gas emissions—such as CO_2, CH_4, and N_2O—and global climate change, pressure is mounting on the international community to take measures to limit and reduce these emissions. In recent decades the international community has made dedicated efforts to address global climate change, from the appeal to protect the climate system at the First World Climate Conference in 1979, to the United Nations Framework Convention on Climate Change adopted at the United Nations Conference on Environment and Development in 1992, to the adoption of the Kyoto Protocol in 1997. In the course of adjusting their energy strategies and formulating energy policies, many countries have adopted additional measures to address climate change, focusing on restricting fossil fuel consumption and encouraging energy-saving efforts and clean energy use. Indeed, since the world oil crises of the 1970s, the issue of climate change has already established constraints on world energy development and has become a driving force for energy conservation and the development of alternative sources of energy. Nuclear, hydro, wind, solar, biomass, and other low-carbon and carbon-neutral forms of energy are now priorities in many countries' plans for future energy development. In 2006 nuclear energy used in the form of electricity accounted for 5.8 percent of the world's total primary energy consumption,[12] while nuclear power accounted for 14.8 percent of

---

[11]IEA's *World Energy Outlook* 2007.

[12]This conversion of nuclear power energy into primary energy is based on the fact that the average conversion efficiency of modern thermal plants is 38 percent and is calculated on the basis of equivalent heat value.

total electricity consumption worldwide. In the twenty-first century, some countries are once again stressing the development of nuclear power and increasing the proportion of nuclear energy in electricity and primary sources of energy. More than 50 countries have passed laws, regulations, or action plans setting forth clear-cut goals and strategies for developing renewable forms of energy. I believe that the trend toward low-carbon and carbon-neutral energy technologies will keep growing as the international community better understands the importance of reducing greenhouse gas emissions.

e. Internationally, there is a clear trend toward the politicization of energy issues, and factors unrelated to supply and demand are playing a bigger role

At present, more than 70 percent of global energy trade is trade in oil. Since the 1970s, the world oil market has undergone several major fluctuations. Oil-exporting countries, oil-consuming countries, and various international forces are engaged in machinations, with the result that factors unrelated to supply and demand are having an increasing impact on international oil prices. The Middle East and other regions rich in oil and natural gas resources are subject to the influence of major international political, military, and economic events, leading to restrictions on and interference in normal trade and investments in oil and natural gas. The rapid growth of capital markets and the virtual economy, the considerable increase in financial derivatives, and the excessive flow of speculative funds around the globe have all had an effect on the oil market. The importance of these external factors makes it more difficult for some developing countries to protect their interests and adds variables to international oil and natural gas exploitation, pipeline construction, market supply, and normal corporate mergers and acquisitions. In recent years, while generally rising overall, international oil prices have continued to fluctuate. This uncertainty is not only the result of changes in supply and demand relations, oil trading in international financial markets,[13] and exchange rate fluctuations,[14] but also of geopolitics, policies of the big powers, public expectations, public opinion, and various emergencies.

---

[13]Oil has increasingly become a means of investment and speculation in international financial markets.

[14]Changes in the exchange rate affect oil prices because international oil trades mostly quote and settle in U.S. dollars, and devaluation of the dollar or fluctuations in the dollar exchange rate affect international oil prices.

2. China's current energy development

a. China has a great variety of energy resources but a low per capita share of them

China has many kinds of energy resources, including abundant amounts of hydro energy and coal. The country's hydro energy reserves rank first in the world, and its coal reserves rank third; however, there are insufficient high-quality reserves of other fossil-based resources. Indeed, in terms of proven remaining economically recoverable reserves[15] of oil and natural gas resources, China currently ranks only thirteenth and seventeenth in the world respectively. Due to its huge population, China's per capita share of all energy resources is below the world average (see Figure 10). Economically recoverable hydro energy reserves total 402 GW, with a total annual power output of 1.75 trillion kW·h. These resources are distributed mainly in the Southwest and are largely underdeveloped; further development would be costly and technologically challenging. Proven remaining recoverable coal reserves are 184.2 billion tons; however, as these are distributed mostly in arid central and western regions far from consumption centers, the general conditions for their exploitation are unfavorable. Proven remaining economically recoverable oil reserves total only 2.04 billion tons,

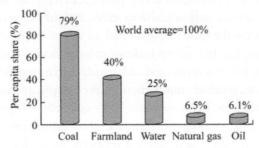

Figure 10   China's per capita share of major resources as a percentage of world averages[16]

---

[15]The cumulative, proven, economically recoverable reserves minus the portion exploited. The economically recoverable reserves are the reserves geological surveys prove are available for economic exploitation under certain technological conditions. This concept is used mainly in reference to oil and natural gas.

[16]Statistics from the Ministry of Land and Resources of the People's Republic of China.

and the reserves-to-production ratio is low. Although there is potential to find more oil reserves, the resulting production increases would be modest. Proven remaining economically recoverable natural gas reserves amount to 2.39 trillion cubic meters, with considerable potential for additional reserves and strong possibilities for substantial increases in production. However, total reserves and exploitation conditions for natural gas in China are not comparable with those in resource-abundant countries such as Russia and Iran. Renewable energy resources such as wind and solar energy are abundant in China, but their development and utilization will be largely determined by technological and economic factors.

b. Although efforts to develop energy are continually being strengthened, energy efficiency remains low

Since the 1990s China's total output of primary energy has more than doubled, reaching 2.37 billion TCE in 2007, making it the world's second largest energy producer. The electric power industry has developed by leaps and bounds. By the end of 2007, total installed capacity exceeded 700 GW; advanced equipment such as 1 GW ultra-supercritical thermal power units and 700 MW hydro turbo-generators were being produced domestically; a number of large, modern coal mines had gone into operation; and new breakthroughs had been achieved in the exploration and exploitation of oil and natural gas. The country had also made good progress in conserving energy and reducing consumption. In the last two decades of the twentieth century, China quadrupled its total economic output while only doubling its energy consumption, yielding an energy consumption elasticity coefficient of 0.43. However, it must be noted that China's energy efficiency remains low and its energy production and use still primitive. Energy consumption per unit of GDP rose between 2003 and 2005, and since 2006 energy-saving and emission-reduction measures have been strengthened. As a result, energy consumption per unit of GDP has decreased, but still greater efforts are needed to sustain this downward trend.

c. The rapid increase in energy production puts considerable pressure on the ecosystem

Driven by ever-growing demand, China's energy production, particularly coal production, has increased rapidly. Over the past 6 years China's annual raw coal output increased by nearly 1.2 billion tons, reaching

2.54 billion tons in 2007 and accounting for approximately 40 percent of the world's total. Nevertheless, a number of problems have accompanied the mass production and utilization of coal, including low recovery rates, substantial waste, frequent accidents, high death rates, and severe damage to the land surface ecosystem and water table. Moreover, emissions of SO_2, smoke, particulate matter (see Figure 11), NO_x, and CO_2 climbed somewhat, posing a new challenge to the management and control of the ecosystem. China is a developing country, and its per capita CO_2 emissions are lower than the world average, but it also faces pressure to reduce greenhouse gas emissions (see Figures 12 and 13).

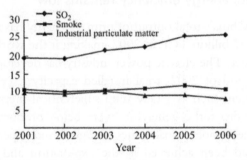

Figure 11   China's main pollution emissions (in millions of tons) [2]

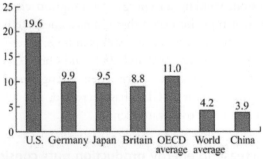

Figure 12   Comparison of international per capita CO_2 emissions (in tons)[17]

---

[17]IEA's *World Energy Outlook* 2007.

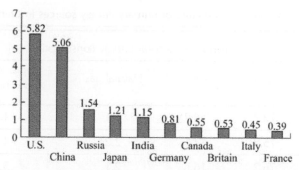

Figure 13   Major countries' CO_2 emissions from fuel combustion (in billions of tons)[18]

d. The coal-intensive energy structure needs to be optimized

Since market-oriented reform and the opening up of China, and especially since the 1990s, China's energy structure has been evolving in the direction of optimization (see Table 1). The proportion of coal in total energy consumption dropped from 76.2 percent in 1990 to 66.3 percent in 2002, but it has increased in recent years, reaching 69.4 percent in 2006. By contrast, the average proportion for developed countries is only about 21 percent. China is the world's largest coal producer and consumer. In primary energy consumption, China's coal consumption is 41 percent points higher than the world average, whereas its oil and natural gas consumption is 36 percent points lower and its hydro and nuclear power consumption 5 percent points lower. At present, clean and renewable forms of energy are still insufficiently developed and used; with the development of wind, solar, and biomass energy still in the initial stages, the task of adjusting and optimizing the energy structure is formidable.

e. Energy demand continues to increase, and sustainable development faces challenges

As China's economy undergoes sustained and rapid development, the country's industrialization and urbanization are accelerating, patterns of consumption are changing, and energy demand is constantly rising. As a result, for the foreseeable future it will be difficult to lower the energy consumption elasticity coefficient substantially. Moreover, the growth of oil and natural gas demand will outpace that of coal. However, restraints

---

[18]IEA's *CO_2 Emissions from Fuel Combustion* 2005.

Table 1   Consumption structure of primary energy sources in China [5]

Year	Proportion of total energy consumption (%)			
	Coal	Oil	Natural gas	Hydro, nuclear, and wind power
1980	72.2	20.7	3.1	4.0
1985	75.8	17.1	2.2	4.9
1990	76.2	16.6	2.1	5.1
1995	74.6	17.5	1.8	6.1
2000	67.8	23.2	2.4	6.7
2001	66.7	22.9	2.6	7.9
2002	66.3	23.4	2.6	7.7
2003	68.4	22.2	2.6	6.8
2004	68.0	22.3	2.6	7.1
2005	69.1	21.0	2.8	7.1
2006	69.4	20.4	3.0	7.2

imposed by natural conditions will make it difficult for domestic supply to keep up. In 2006 China's oil reserves-to-production ratio was only 11.1, far below the world average of 40.5 (see Figure 14). The imbalance between the country's energy supply and demand—oil and natural gas in particular—will therefore become more pronounced. As a result, only if China increases its efforts to save energy, accelerates industrial restructuring, and sensibly guides consumer behavior will it be able to gradually slow increases in energy demand and eventually achieve low or zero growth in demand for fossil fuel-based energy.

III. STRATEGIC THINKING ON CHINA'S ENERGY DEVELOPMENT

The coming decades will be a crucial period for China's comprehensive economic and social development and for the Chinese nation's great rejuvenation, and the task of energy development will be difficult. To meet the

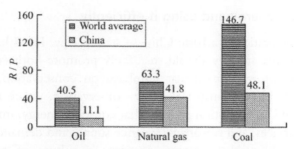

Figure 14 Comparison between China and the world of the reserves-to-production (R/P) ratios of fossil fuel energy resources in 2006[19]

ever-increasing energy demands of over one billion people while building a moderately prosperous society in all respects, in the next 20 years China needs to establish the world's largest energy supply and consumption system. The country thus faces the urgent task of blazing a new and distinctively Chinese trail in energy development in order to achieve its strategic goal of modernization at the lowest possible cost to its energy resources and the environment. The energy system is enormous, and it will take several decades to restructure it and completely upgrade the current generation of technologies and facilities. Energy development therefore requires long-term strategic thinking and a search for the optimal development path.

1. The strategic orientation for energy development

In contemplating China's energy development strategy, we need to adopt a broader and more open-minded perspective, give energy a prominent place in our national development strategy, acquire a keen understanding of energy development trends, and rapidly improve the goals, principles and tasks of our energy strategy. Pursuing a new and distinctively Chinese energy development path means steadfastly conserving energy, using it efficiently, diversifying development, keeping the environment clean, becoming technology-driven, pursuing international cooperation, and committing to the creation of a system of energy production, distribution, and consumption that is highly efficient, technologically advanced, low in pollutants, minimally threatening to the ecosystem, and above all effective in providing a steady and secure energy supply.

---

[19]Statistics from the Ministry of Land and Resources of the People's Republic of China 2007, and other sources.

a. Conserving energy and using it efficiently

Resource conservation is a basic Chinese state policy. For a long time to come, our energy strategy should steadfastly promote both conservation and development, with particular emphasis on conservation. We must resolve to give high priority to energy conservation, create methods of exploitation and consumption patterns that conserve energy, improve general energy services, and reasonably balance supply and demand. We need to advocate ecological awareness, promote a culture of conservation through energy education, and leverage technological advances. We need to increase significantly the efficiency of the country's energy system and make China's energy-intensive industries as functional as those of the rest of the world as quickly as possible. We must constantly raise overall energy efficiency and support maximum economic and social development with minimal consumption of energy resources.

b. Diversifying development

Only by fully using all the energy resources that exist in significant quantities in China can we optimize our energy structure and satisfy future energy demands. Developed countries have already optimized their use of energy from fossil fuels and are now developing low-carbon energy and moving toward greater optimization of energy resources. China also needs to diversify energy development, accelerate energy restructuring, increase oil supplies, and significantly raise the proportion of natural gas, nuclear energy, and renewable energy in energy production and consumption. In our quest for new energy supplies we must strive to rely primarily on high-efficiency, clean, new, and low-carbon or carbon-neutral renewable forms of energy.

c. Keeping the environment clean

Controlling pollution, protecting the environment, and alleviating ecological pressure are important requirements for China's energy development. In light of these needs, our energy strategy should consider how it can most effectively address the challenges posed by global climate change. To solve environmental problems arising from energy consumption, we must take measures to increase the use of clean energy, work toward the development of environmentally friendly forms of energy, use coal more cleanly and efficiently, and promote clean use of energy in industry, transportation, and buildings to minimize the pollution emissions and ecological damage caused by energy production and consumption. In this way we can develop and utilize energy while protecting the ecosystem.

d. Becoming technology-driven

Energy development needs to be driven by technology. The only way we can constantly improve energy efficiency, develop clean energy, achieve sustainable energy development, and press ahead with modernization is through constant technological innovation. Looking to the future, we must employ the most advanced energy technology available, plan ahead for research and development in energy technology, and establish a reserve of energy technology. Energy production and conversion technologies are constantly improving worldwide, facilities are becoming larger and taking advantage of economies of scale, and energy production is becoming more capital intensive and highly concentrated. China's energy industry must follow a path of intensive development, improve its scientific and technological innovation capabilities, and increase its international competitiveness.

e. Pursuing international cooperation

The solution to China's energy problem has great significance for the world. By enhancing international energy cooperation, we can promote economic and technological exchanges in the energy field and create more channels for opening up the energy sector to the outside world. Chinese enterprises must "go global" and expand investment in overseas markets to develop more energy resources and improve oil and natural gas supply capabilities. Increased contacts with other countries on energy issues can facilitate dialogue and coordination on strategies and policies and promote consistent improvement of the global energy security system. Such improvement will help to increase not only China's energy supply but also the world's.

2. Long-term strategy of giving high priority to energy conservation

a. Making energy conservation a high priority is in accordance with China's basic national conditions

At present, our country's per capita energy consumption is low, but economic and social development will cause it to increase in the future, and the total volume will continue to expand. However, China cannot blindly follow the traditional development model of developed countries and rely on greatly depleting the world's energy to fuel a high level of energy consumption. We need to make great efforts to find a new development path; consistently make energy conservation a high priority; and achieve industrialization, urbanization, and modernization through energy-efficient development.

b. Energy conservation is an important requirement for balancing energy supply and demand

China must make energy conservation a high priority in order to balance its future energy supply and demand. Together, strengthening energy conservation and promoting energy efficiency can effectively stop energy demand from growing too rapidly, keep our country's total energy demands within the limits imposed by our resources and environment, and achieve economic and social development with high efficiency and low consumption.

c. Energy conservation is a tangible manifestation of modern civilization

Thrift is a traditional virtue of the Chinese nation, and cherishing resources and protecting the environment are important indicators of modern civilization. Therefore, we need to guide our entire society to cultivate a mindset of consuming energy sparingly, establish rational consumption patterns, and encourage reasonable and appropriate consumption behavior. We need to build efficient and energy-saving public facilities as well as improve our resource allocation system and mechanisms to promote energy conservation. Furthermore, we need to conserve energy resources and improve energy efficiency in all areas throughout our economic and social development.

d. Industry, transportation, and buildings are key target areas for energy conservation

Improving energy efficiency is an important step in implementing the strategy of giving energy conservation high priority. We need to reduce unreasonable energy demand as far as possible and use energy more efficiently in order to provide more and better energy services with relatively little investment in resources. Industry, transportation, and buildings are key areas for energy conservation.

(i)   *Energy conservation in industry.* Industrial energy consumption accounts for a large proportion of China's total consumption, and the huge potential for savings should be fully tapped. As our industrialization continues, we need to take advantage of our position as a late starter and widely adopt advanced processes and technologies to bring our industrial energy consumption up to the world's most advanced level. Globally, the pace of technological advances is increasing, and new procedures and equipment are constantly emerging. The focus of raising energy efficiency is shifting from the technological refinement of individual appliances to system optimization and holistic improvement. Current overall energy consumption by the world's leading iron and steel companies is only 630 kg of standard coal per ton of steel, and this may eventually fall to below 570 kg. The per-unit consumption of a large-scale advanced dry-process cement rotating kiln is only 96 kg of standard coal equivalent, and this may drop to below 86 kg in the future.

(ii) *Energy conservation in transportation.* In industrialized countries, the energy consumed in transportation accounts for 30 percent to 40 percent of the total. At present, the proportion in China is still relatively low, but as more and more families buy cars, the amount of energy the transportation sector consumes will rise dramatically, and we must pay careful attention to this trend. The public transportation system can offer convenient and energy-efficient transportation for the general public, both reducing traffic jams and benefitting the environment. In Tokyo, for instance, 80 percent of passenger trips are taken by public transportation, 70 percent of them by rail. Making the transportation system and the means of transportation more energy-efficient and environmentally friendly has already become an international trend. Take the current gasoline-electric hybrid cars for example: gasoline consumption can be as low as 3 to 3.5 liters per 100 km, 50 percent to 70 percent lower than cars powered by traditional means, and there is potential for further reduction. Some countries are in the process of developing cars powered solely by electricity that can run 500 km on a single charge, and demonstration models of fuel-cell vehicles powered by hydrogen energy with zero CO_2 emissions are beginning to appear. In the course of our country's industrialization and urbanization, oil-saving and environmentally friendly goals for the development of the automotive industry should be given high priority in order to bring the industry's energy efficiency gradually into line with leading levels elsewhere in the world. At the same time we must energetically develop high-speed commuter and intercity rail systems to streamline public transportation significantly and thus reduce energy consumption and pollution.

(iii) *Energy conservation in buildings.* Energy is consumed in buildings for heating, air conditioning, ventilation, lighting, hot water, elevators, and office and home appliances. In industrialized countries, energy consumption in buildings accounts for more than 30 percent of the total. China is currently in the process of rapidly developing its construction industry, adding approximately 2 billion m^2 of floor space each year. In fact, China is now the largest construction market in the world, and its energy consumption in buildings is gradually increasing. Therefore, government office buildings and public facilities need to take the lead in promoting energy conservation and thus act as models for residential and commercial buildings. We must actively push for the application of energy-saving technologies to achieve significant reductions in the energy demands of buildings—for example, high-efficiency insulation materials; low heat emission glass; high-efficiency heating and cooling systems; solar hot water systems; water, ground, and air source heat pumps; energy-saving lighting; and more intelligent buildings. Such innovations will reduce energy consumption in new buildings by 50 percent to 65 percent, and buildings with extra-low energy consumption will save 90 percent. In the future it may be possible to construct new buildings with low or even zero CO_2 emissions. We can carry out comprehensive technological upgrading to streamline central heating systems by improving terminals, pipeline systems, and heating sources, thus increasing the current efficiency of less than 55 percent to approximately 85 percent [6]. We need to strengthen energy conservation standards for buildings, promote energy-saving building design, and use energy-saving materials and equipment in buildings to continue to improve the operating efficiency of their energy systems. In this way we can effectively reduce energy demand growth in buildings while at the same time making people's homes and living conditions better.

3. Effective development and use of primary energy sources

China's total energy demand is huge, and the only effective way to guarantee sufficient energy for economic and social development is by constantly increasing the energy supply capacity. However, we can meet the total demand for different amounts and varieties of energy only by making full use of multiple energy resources. Coal is expected to remain an important primary energy source, but its consumption as a proportion of overall use will diminish. Conversely, consumption of oil, natural gas, nuclear energy, and renewable types of energy such as hydro, wind, and solar energy will increase, gradually forming a diversified and optimized energy structure.

a. Coal

Coal is China's most important energy source. The amount of coal produced and consumed in China is huge, both in quantity and proportion, making it difficult to replace. Therefore, the development of safe, efficient, clean, environmentally friendly, and sustainable coal production is of great significance. To raise China's coal production to the world's most advanced standards as quickly as possible (see Table 2), coal mining should

Table 2 Technical and economic indices in the world's major coal-producing countries[20]

Country	Production (in millions of tons)	Staff	Productivity (tons/man-shift)	Mining mechanization ratio (%)	Overall mechanization ratio (%)
United States	993	71,700	13,849	100	51.3
Australia	350	21,200	16,509	100	100
South Africa	215	42,500	5,059	97	97
Germany	211	48,700	4,333	100	100
Poland	162	140,000	1,157	99.2	95.3
China	2,190	5,500,000	398	45	30
Key state-owned mines in China	1,040	2,580,000	403	83	72

[20]State Administration of Coal Mine Safety and National Bureau of Statistics of China, Department of Energy of the United States, and IEA. The data for China is from 2005, and for other countries from 2004.

be based on safe production, scientific and technological progress, and better management. In order to achieve sustainable exploitation of coal and modernize the coal industry, we must establish a number of large coal bases and coal enterprises, as well as modern mines with leading technology, good security, mechanized mining, and high production efficiency. We must also give consideration to the ecosystem throughout the process of coal resource development. We must make rational decisions regarding the scale of mining in accordance with different local ecosystems. We must develop a circular economy, comprehensively treating and using coal tailings, mine water, and coal powder. We must clean up and restore the ecosystems of exhausted mines and effectively protect the ecosystems of mining areas and their surroundings.

In developing coal, we also need to address the issues of efficient and clean use. That means considering not only energy conversion efficiency but also the control and reduction of SO_2 and NO_x emissions. In addition to developing clean coal technologies such as circulating fluidized bed combustion and large-scale gasification and improving direct and indirect liquefaction technologies, we also need to accelerate research and development for the newest generation of clean coal technology: coal polygeneration (see Figure 15). Polygeneration based on coal gasification can simultaneously generate electricity, thermal power/steam, liquid fuels, and chemical products. It can also achieve near zero emissions of SO_2, NO_x, particulate matter, microelements, and organic substances. Furthermore, removing CO_2 in polygeneration is easier than capturing it from smokestacks after coal is burned, which makes coal polygeneration an important measure in reducing future greenhouse gas emissions. Compared with the production of electricity, liquid fuels, and chemical

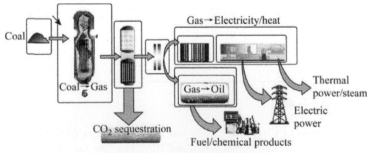

Figure 15   Schematic diagram of coal polygeneration [7]

products, coal polygeneration can improve overall fuel efficiency by 10 percent to 20 percent and significantly reduce investment per unit of production. The United States, Japan, and some European countries have already designated coal polygeneration as a new area of clean coal technology and have formulated detailed research and development plans. Because China is a particularly large supplier and consumer of coal, the research, development, and application of coal polygeneration technology in this country holds special significance.

b. Natural gas

With high calorific value and low CO_2 emissions, natural gas is a clean fossil fuel and a good choice for meeting energy demand in cities and other densely populated regions; it should therefore be developed and used on a large scale. The development of natural gas can effectively relieve environmental pressure, greatly improve the efficiency of the energy system, and stimulate the development of a variety of advanced high-efficiency energy technologies and application systems—including high-efficiency power generation, distributed energy systems, and clean transportation fuels. With the maturation of long-distance and large-scale natural gas pipeline technology and the universal adoption of transmission and utilization technologies for liquefied natural gas, the exploitation and use of natural gas have expanded rapidly across the world. In recent years, China's proven reserves of natural gas have quickly risen, offering great potential for increased production. Coal bed gas resources are considerable, and their application technology is constantly improving. Moreover, future imports of foreign natural gas and liquefied natural gas resources look promising. China has many large cities and densely populated regions where demand for natural gas is great, so conditions for its large-scale and rapid development are favorable. If we can effectively increase the natural gas supply and substantially increase its proportion in total energy consumption, we will be able to optimize our country's energy structure.

c. Oil

Oil is a high-quality energy source with many outstanding characteristics, and oil products are still the superior choice for transportation fuel. With the growth of per capita income and the acceleration of industrialization and urbanization, car ownership in China has risen sharply, as has the number of other vehicles on the road, boosting demand for liquid fuels. However, we can meet the growing need for oil only through domestic development and expanding oil imports. Our country still has huge potential oil and natural gas resources both onshore and offshore, and the maintenance of a long-term stable oil output with moderate increases in production is vital to ensuring energy security. In order to increase domestic oil output, we need to guarantee our investment, intensify our geological surveying, and strive to achieve breakthroughs in surveying technology and geological theory. After a long period of exploration, the continental hydrocarbon generation theory has already become a part of the theoretical system of petroleum geology, but the oceanic hydrocarbon generation[21] theory still needs to be established and refined. We still need to improve the oil and natural gas resource assessment system, as well as raise the standards of equipment and technology for use in low-permeability oil-gas reservoirs[22] and deep-sea oil-gas surveying and exploitation. The distribution of oil resources in the world is relatively concentrated; therefore, we should actively pursue international cooperation and use more overseas resources, among other channels, to add to our oil supply capacity.

d. Renewable energy sources

The world's renewable energy resources have huge potential, and everyone has great expectations for them. Accelerating the development of renewable energy can help find long-term strategic energy substitutes and strengthen environmental protection. China has already enacted a law that makes the development and use of renewable energy a priority in energy development, and it has also drawn up a long- and medium-term development program

---

[21]The general term for basin formation, hydrocarbon accumulation, and reservoir formation in oceanic deposition environments.

[22]Low-permeability oil-gas reservoirs are perforated and permeable geological bodies with a certain amount of oil or natural gas concentrated therein and covered by nonpermeable rock strata. The permeation rate differs with the size of rock cracks and the growth of connectivity. According to international standards, low-permeability oil-gas reservoirs are oil reservoirs with a permeation rate of lower than 50 millidarcy (mD) (In a substance having a permeability of 1 mD, a fluid with a viscosity of 1 mPa·s under a pressure differential of 101.325 kPa [1 atm] flows 1 cm per second through a core with a surface area of 1 cm^2, i.e., flows at a rate of 1 cm^3 per second.)

Table 3 Major targets for the long- and medium-term development program
for renewable energy [8]

Year	Hydropower (GW)	Biomass power (GW)	Wind power (GW)	Proportion of renewable energy in primary energy consumption (%)
2005	117	2.2	1.26	7.5
2010	190	5.5	5.0	10
2020	300	30	30	15

for renewable energy (see Table 3). Given the many kinds of renewable energy
and the varying conditions of resources, technological maturity, and economic
feasibility, we must carry out the development and utilization of renewable
energy with consideration for local conditions, guidance tailored to specific conditions, and a clear focus.

Hydro energy is an important renewable energy. As the technology for
its development and utilization is already mature, it is the focus of development in the near future. Wind power is another abundant resource
whose utilization technology is basically mature. It could therefore be a
priority for current large-scale development in order to build a significant
supply capacity. Solar energy resources offer great potential, and we should
be able to apply them extensively as soon as further breakthroughs in the
key technology are achieved and their cost-effectiveness improves. We
need to develop solar power generation technology and heat utilization
technology and promote the use of solar water heaters in combination
with energy conservation in buildings.

Internationally, the development and utilization of biomass energy have
received considerable attention, and some countries with abundant land
resources and excellent conditions for growing crops have already identi-
fied biomass energy as an important energy substitute. For example, the
United States is developing the corn–based fuel ethanol on a massive scale,
while Brazil is focusing on developing ethanol from sugarcane. The ques-
tion of whether mass fuel production from crops will lead to global
shortages of grain supplies has become a topic of international debate. Each
country's conditions vary, and the decision to develop and use biomass

energy must be based on a country's specific conditions. China's biomass energy comes mainly from agriculture, forestry, livestock, and industrial and municipal waste, along with some farm and forest crops. But although China has many biomass sources, they are scattered and costly to collect. Nevertheless, they are suitable for developing various utilizing technologies and for use in a number of consumer markets. Many factors, including scarce land resources, limited farmland, and a fragile ecosystem, restrict the production of fuel from farm crops. In developing biomass energy, China must first use existing waste resources and gradually increase the level of their exploitation and use. Some mature technologies, such as rural marsh gas and power generation from straw, should also be actively promoted and supported. In addition, we need to invest a reasonable amount in research of other key technologies on the verge of a breakthrough, such as cellulose ethanol.

e. New forms of energy

The development of new sources of energy and related technologies is of great significance for future energy substitutes. We need to give this issue our full attention, promptly make preparations to develop them, actively follow the lead of other countries, and organize research and development initiatives. New energy development and utilization must rely on advanced energy technologies. Although research into hydrogen energy, gas hydrates, and nuclear fusion has not yet translated into practical production capabilities, the rapid progress of science and technology gives people hope for the future of new energy.

Hydrogen energy has a promising future and may emerge as a practical new type of clean fuel. Hydrogen generated via various methods can be directly converted into electricity by fuel cells, and when used in cars, trains, and other means of transportation it produces zero emissions of pollutants. It can also be used in fixed power generation or heating facilities in industrial, commercial, or residential buildings (see Figure 16). A number of major industrialized countries have made some progress in developing cars powered by hydrogen fuel cells. Once key technologies have been developed for inexpensive hydrogen generation using nonfossil-based forms of energy, safe hydrogen storage and transportation, and high-efficiency durable fuel cells, (especially with large-scale hydrogen generation by solar, nuclear, or biomass energy) hydrogen energy use should become practicable.

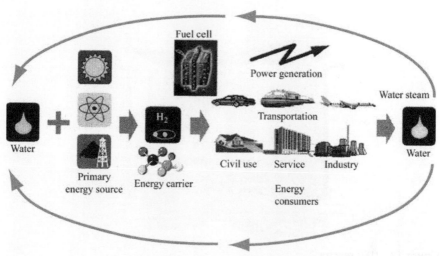

Figure 16   Schematic diagram of hydrogen energy utilization[23]

Gas hydrates have been called "the potential new energy of the twenty-first century" and have drawn extensive attention from scientists and governments in many countries. Also known as solid methane, gas hydrates consist mainly of methane and water molecules in a solid state. This special type of unconventional natural gas resource is abundant in the ocean floor around continents and underground in permafrost zones. According to a report by the United Nations Intergovernmental Panel on Climate Change, gas hydrate reserves may be twice as extensive as the combined total of all other fossil-based energy resources, and the estimated recoverable amount may equal total known oil and natural gas reserves. However, a number of difficulties stand in the way of the practical exploitation and use of gas hydrates. Nevertheless, China has already listed technology for exploiting gas hydrate energy in its National Program for Long- and Medium-Term Scientific and Technological Development and plans to conduct an in-depth resource survey, undertake applied research, and seek out technology for exploiting gas hydrates.

The world is placing great hope in thermonuclear fusion energy. The earth's usable fusion materials are enormous, and once the technology for controlling thermonuclear fusion succeeds, it will open a new chapter

---

[23]Discussion material accompanying China's National Program for Long- and Medium-Term Scientific and Technological Development.

Figure 17   ITER reactor[24]

in energy applications. The International Thermonuclear Experimental Reactor (ITER) project has already been launched, and if it progresses smoothly, demonstration of commercial power generation by controlled thermonuclear fusion may become a reality by 2050 (see Figure 17). ITER seeks to control thermonuclear fusion by means of magnetic confinement, but there are also other approaches, such as inertial confinement. While actively participating in ITER, China also needs to explore and research these alternative technologies. In addition, there are choices regarding fusion materials. Currently, most experts favor the reaction between deuterium and tritium (isotopes of hydrogen), whereas others favor the reaction between deuterium and helium-3, an isotope of helium.

4. Development of an advanced electric power system

Electric power holds a particularly important position as an energy source. Electric power technology makes it possible to convert primary energy from fossil fuels—as well as from nuclear, hydro, wind, and solar energy—into highly efficient, clean secondary energy that can be easily transmitted and used and can offer quality end-user services, improve system efficiency,

---

[24]http://www.iter.org.

and control environmental pollution. The long-term trend in global electric power technology is to increase the efficiency of energy conversion and transmission and achieve highly efficient, clean power generation and power supply security.

Over the past 2 years China's electric power industry has grown rapidly, and the approximately 100 GW of new installations a year have basically eliminated the shortage of electric power supplies. In 2007 the installed capacity of electric power reached 713 GW (see Figures 18 and 19). Of this amount, the installed capacity of hydropower amounted to 145 GW. The Three Gorges Power Plant has put 21 power units into

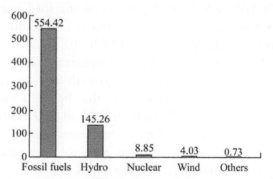

Figure 18   China's total installed electric power capacity in 2007 (GW)[25]

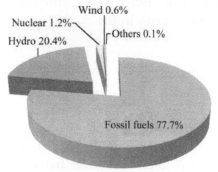

Figure 19   China's electric power distri-bution in 2007[26]

---

[25]National Bureau of Statistics of the People's Republic of China, and the China Electricity Council.

[26]*Ibid.*

operation, and construction is under way on large hydropower plants in Longtan, Xiangjiaba, and Xiluodu. When two power units in the phase one construction of the Tianwan Nuclear Power Plant went into operation, the installed capacity of nuclear energy reached 8850 MW. The development of wind power has speeded up, and there is now an installed capacity of 4030 MW. Power generation equipment technology continues to improve, and supercritical and ultra-supercritical power units are now being mass-produced and installed. Obsolete power units are being phased out more quickly, and more than 20 GW of small thermal power units have been shut down over the past 2 years. Desulfurization equipment has already been installed on coal power units with a generating capacity of 270 GW, which is 45 percent of the total installed capacity of thermal power plants. In 2007 the amount of coal consumed in supplying electric power declined to 357 g/kW·h, 10 g/kW·h lower than in 2006. It is predicted that large-scale development of electric power will continue in the near future; that the growth rate of electric power will soon exceed that of primary energy; that by 2010 installed electric power capacity in China will reach or exceed 900 GW; and that in 5 years or so, China is likely to have the largest electric power system in the world. As we move toward these goals we must raise the standard of the electric power industry, develop clean and highly efficient coal power, increase the proportion of hydro and nuclear energy we use, promote wind and solar power generation, and enhance power transmission and conversion to form a diversified power structure and one of the most advanced electric power systems in the world.

a. Thermal power

A high proportion of China's energy comes from coal, which is mostly converted into electricity for end-user consumption. Economic and social development trends suggest that the proportion of coal converted into electric power will continue to rise in the future, that power generation facilities and energy efficiency will further improve, and that emissions from coal burning will be concentrated in the electric power industry. Strengthening environmental protection requires the development of highly efficient, clean coal power generation technology, exploration of low-carbon power generation technology, and improvements in technology and equipment to control SO_2 and NO_x emissions. With the use of supercritical and ultra-supercritical power units, fuel efficiency can reach 41 percent to 43 percent and 45 percent to 47 percent respectively. Research and development of power units with higher temperature and pressure parameters is currently under way. With integrated gasification combined cycle technology, fuel efficiency can reach 50 percent or higher, while also reducing pollution emissions and providing a technological foundation for CO_2 sequestration. Combined cycle gas turbine technology for burning natural gas can further raise fuel efficiency to 58 percent to 65 percent, while the fuel efficiency of distributed combined heating and power generation systems can reach 80 percent or more. These advanced technologies should be developed and put to use quickly, so that China's power generation efficiency can soon reach the level of the world's most advanced countries and keep pace with the new international trend of sustainable development.

b. Hydropower

Hydropower is an important renewable energy source. Most developed countries have already made the use of hydropower resources a priority and have fully developed them. China is endowed with abundant hydropower resources, though two-thirds of commercially viable hydro energy resources have yet to be developed. These clean and renewable energy resources are a very precious asset that ought to be utilized first. Full exploitation of hydropower resources can play an important role in increasing our country's energy supply capacity and optimizing our energy structure. However, the hydropower construction cycle is long, and hydroelectric power plants are usually far from where electricity is needed most. Therefore, we must make the development of these electric power facilities a priority by beginning the preliminary work and preparations as early as possible. We need to pay careful attention to the impact of hydroelectric power plants on the ecosystem and analyze and evaluate related issues in a scientific and systematic manner. Hydropower must be developed systematically in parallel with protecting the ecosystem so that hydro energy resources can be fully utilized. We need to take advantage of the low cost and high performance of hydropower to devise reasonable compensation and support mechanisms and adopt necessary measures to solve the problems associated with relocating people displaced by dam construction. In short, we need to do everything we can to reduce the negative effects of exploiting this form of energy and use all of China's commercially viable hydropower resources.

c. Nuclear power

The world has already made great progress in the use of nuclear energy. With the extensive application of second-generation nuclear power technologies, nuclear power now accounts for a significant share of primary energy (see Table 4). Many countries are developing a new generation of nuclear power technology and working to improve its safety and financial viability even further. Some third-generation nuclear power technologies are being developed in the direction of passive safety[27] and thus

Table 4 Share of nuclear power in primary energy of some countries and regions in 2006[28]

Country (region)	Share (%)	Country (region)	Share (%)
France	38.9	Germany	11.5
Sweden	32.6	Spain	9.3
Switzerland	21.6	United States	8.1
Finland	19.6	Taiwan, China	7.9
Belgium and Luxemburg	14.8	Britain	7.5
Ukraine	14.8	Canada	6.9
Republic of Korea	14.9	Russia	5.0
Japan	13.2	China	0.8
Average of 25 European Union countries	12.7	World average	5.8

---

[27]Passive nuclear safety is a safety feature of nuclear reactors that relies on the laws of physics, the properties of materials or stored energy such as gravity, buoyancy, natural circulation, and energy in compressed gas, instead of on an AC power supply or equipment such as pumps or emergency diesel engines to keep the nuclear plant's emergency safety system functioning for long periods. Passive safety systems increase the level of safety by reducing the risk resulting either from failure by operating personnel to carry out emergency procedures or from the loss of power.

[28]*BP Statistical Review of World Energy* 2007.

simplifying equipment, while others are being developed with a view to increasing redundant safety in relevant facilities; both approaches substantially improve safety.

Third-generation nuclear power technologies have already reached the stage of demonstrating their commercial application. High-temperature gas-cooled reactors (see Figure 20) and fast reactors [9], which may become fourth-generation nuclear power technologies, are also under development. Given that the technology for treating nuclear waste is improving, nuclear power is now a highly efficient and clean type of energy, making it one of the most important options for China to increase its energy supply, optimize the energy structure, and address climate change. Therefore, scaling up nuclear power use should be a focus of our energy strategy. In the long run, nuclear power may reach the scale of 100 GW in China and become a major primary and power generation energy. After years of practical experience, our country has mastered mature and reliable nuclear power technologies and is now in a position to accelerate nuclear power development. To achieve this goal we must actively promote the introduction and assimilation of third–generation

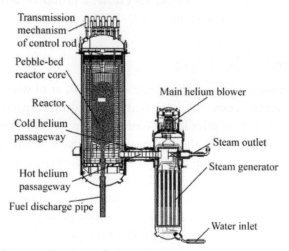

Figure 20 Schematic diagram of the primary-loop system in a high-temperature gas-cooled reactor[29]

[29]Materials from the Institute of Nuclear and New Energy Technology of Tsinghua University, People's Republic of China.

nuclear power technologies, speed up domestic research into advanced nuclear power technologies, master advanced pressurized water reactor technology, achieve standardized scale development of reactors, and raise the level of domestically produced nuclear power facilities. At the same time we should vigorously carry out independent research and development on fourth-generation nuclear power technologies and develop high-temperature gas-cooled reactor and fast reactor capabilities to reach the preliminary commercialization stage as soon as possible. Nuclear safety is the lifeblood of nuclear power development, and therefore we cannot afford the slightest negligence in this regard. To develop nuclear power on a large scale, we must further improve safety management and strengthen security and quality in every aspect of the design, manufacturing, construction, and operation of nuclear power plants. Establishing nuclear energy on a large scale also requires us to develop a corresponding uranium enrichment capability, accelerate the construction of nuclear fuel treatment facilities, and complete a nuclear fuel recycling system to ensure that we have adequate resources to take advantage of nuclear power. Above all, large-scale development of nuclear power requires early training of personnel to create a group of professionals of the highest caliber to work in this field.

d. Wind power

The world is moving toward large-scale development of wind power generation. In recent years the electricity-generating capacity of newly installed wind power facilities in the United States and European countries has become second only to that of natural gas power. Furthermore, wind power technology is developing in the direction of large individual wind turbines (see Table 5). China is endowed with abundant wind energy resources. According to preliminary estimates, the total amount of technically usable wind power on Chinese land could be as much as 1 TW, so the potential for development and utilization is huge. As the use of wind power grows in scale and proportion of domestic energy production, the cost of its generation will significantly decrease, gradually making it economically competitive and thus encouraging more extensive development. While the wind power industry is in its initial stage, we need to provide policy support and cultivate a market for it. We need to support technological innovation by research institutes and enterprises, as well as

Table 5 Development trends in the latest wind power units in the world [10]

Year	Rated power (kW)	Wind wheel diameter (m)	Wheel hub height (m)	Annual capacity (kW·h)
1980	30	15	30	35,000
1985	80	20	40	95,000
1990	250	30	50	400,000
1995	600	46	78	1,250,000
2000	1,500	70	100	3,500,000
2005	5,000 (offshore)	115	90	17,000,000

improving our overall ability to design MW wind power units or above and solve problems concerning the design and production of key components such as bearings, current transformers, and control systems. Through these efforts we will build a sophisticated, internationally competitive industrial wind power system and propel our country's wind power development to the world's front ranks.

e. Solar and other electric power

China is rich in solar energy, which is one of the renewable forms of energy with the highest potential for exploitation. Solar power generation, particularly photovoltaic (PV) power generation, has been developing rapidly around the world, and its evolution from nongrid-connected to grid-connected generation has already reached a significant scale. However, the current cost of PV power generation is still high. It is hoped that technological advances will substantially reduce generating costs and enable solar power generation to enter a period of large-scale growth. At present we must continue to increase research and development investment in solar power generation technology and strive to achieve the breakthroughs in key technologies that will enable us to build a solid foundation for solar energy to become a major electricity source in the future.

In addition, China has many other renewable energy resources with potential for power generation. Biomass power generation (by direct combustion as well as gasification) could be developed in accordance with local conditions, while geothermal heat energy and ocean energy can also be exploited. We must therefore keep pushing ahead with research and development to apply these technologies.

f. Power transmission and distribution systems

The construction of a power grid is an important aspect of building an electric power system. Our territory is vast, the distance between our sources of power in the west and their eastern distribution destination is great, and as a result the quantity of energy that needs to be transported is huge. These factors place high demands on the development of power transmission and distribution technology. Transmitting electric power from large hydro and thermal power bases requires us to make greater efforts to build high-level and large-volume power transmission capabilities. At present, 1000 kV AC and ±800 kV DC ultra-high voltage power transmission technologies and facilities are in the experimental stage. The natural power of a 1000 kV AC ultra-high voltage power transmission circuit (single circuit) is close to 5 GW, about five times as much as that of a 500 kV power transmission circuit. The natural power of a ±800 kV DC ultra-high voltage power transmission circuit (single circuit) can reach 6.4 GW, 2.1 times that of a ±500 kV power transmission circuit, significantly increasing the efficiency of power transmission and reducing the amount of land used as a transmission corridor. It is therefore important to complete the pilot demonstration projects and environmental impact assessment work. The actual power transmission capacity of circuits is closely related to the dynamic characteristics of a system. Optimizing the power grid structure, improving the inverter side system, taking relay protection measures, using power grid control technologies, and carrying out research and development and then applying advanced wide-area measurement systems and the Flexible AC Transmission System[30] will effectively increase system safety and stability and improve power transmission capacity. Modern social activity depends heavily on an electric power supply; therefore, we must further

---

[30]A kind of AC power grid transmission technology with variable parameters. It can equalize surges to maintain stability and enhance system transmission capacity by adjusting the resistance, capacitance, and inductance of the power grid.

develop safety and stability control technologies for the power grid and improve power grid management to create a safe and highly efficient power transmission and distribution system. Furthermore, given the increased use of discontinuous sources such as wind and solar power, we must also improve the technology and our ability to connect diverse power sources and transmit and distribute the power they supply.

Many new technologies are under development that are closely related to advanced electric power systems, for example, fixed fuel cells, superconducting materials and their application, high-parameter high-power gas power generation, and microturbine power generation.[31] Nowadays, power generation technologies are being developed not only to enable higher parameters and larger production facilities but also to allow distributed microproduction, and we must closely follow and actively participate in both trends. The development of power generation, transmission, and distribution technologies also lends impetus to technological development in new materials, power electronics, automatic controls, advanced manufacturing techniques, and high-grade precision equipment manufacturing. To achieve comprehensive improvements in electric power technology, we must conduct future-oriented research into new electric technology and scientific theory, lay solid foundations, and strive to build up our independent innovation capacity.

IV. ENERGY DEVELOPMENT POLICY

Energy policies are an important prerequisite for the implementation of an energy strategy. The energy policy adopted should fully reflect the intent of the energy strategy and assure the right conditions for a new, distinctively Chinese energy development path. To improve our country's energy development policies, we should emphasize the key role of energy in economic and social development and adopt specific energy policies to ensure that the goals and principles of our energy strategy are implemented. We need to establish systems and mechanisms favorable to sustainable energy development and make sure that the market plays its basic role in energy allocation to the fullest, while strengthening

---

[31]Microturbine power generators are usually gas turbine power generators used for homes or single building units with power of 1 MW or below.

and improving government guidance and management. We need to give substance to the idea that "science and technology are a primary productive force" and encourage increased investment in energy technology that will constantly drive innovation forward. From the perspective of China's long-term development, we need to consider the following eight operating principles in order to implement our energy development policies effectively.

1. Give energy development policies an important place in national economic policies

Energy is the food of industry and the lifeblood of the national economy. Energy issues are more than major economic and social issues; they also involve major diplomatic, environmental, and security issues. Ensuring a stable energy supply while adapting to and slowing climate change is a very challenging strategic issue critical to China's long-term development. National policies concerning macro-control, industrial development, finance and banking, science and technology, and international economic relations should fully emphasize and embody energy conservation and sustainable energy development.

2. Make advance arrangements for major energy construction projects and energy science and technology projects in key fields

Many energy construction projects are immense in scale; span a number of regions and industries; and require long-term planning, advance arrangements, coordinated policies, and effective implementation. Projects such as the national oil and gas pipeline network, national oil reserve bases, hydropower cascade developments in river basins, large coal bases, and power grid systems all require extensive consultation, early planning, and orderly development. For major energy technologies—such as next-generation clean and high-efficiency coal utilization technology, advanced nuclear energy technology, and the use of solar energy—we need to take preventive measures, formulate careful plans and implement them as early as possible, and organize and coordinate our efforts to achieve technological breakthroughs that will create a reserve pool of power technology.

3. Increase funding for energy development and scientific and technological innovation through multiple channels

Innovation in energy technology determines future energy development and as such deserves an important place in national scientific and technological innovation. Energy technology projects are usually large in scale and involve considerable basic research. Therefore, they require significant funding for research and development, verification, demonstration projects, and supplementary projects. In addition to providing increasing financial support from the national treasury, we also need to motivate private enterprises to invest in this work. To build key national projects, we need to get enterprises to play a leading role, attract investments from a broad spectrum of society, and use modern financial methods to broaden the sources of capital while the government continues to provide the necessary fiscal policy support.

4. Establish a rational pricing mechanism for energy resources

Most conventional energy resources, particularly fossil fuels, are nonrenewable. Energy prices should fully reflect the scarcity of resources, the supply and demand situation in the marketplace, and the cost of ecological protection and environmental management. Such pricing would send the right signals to all stakeholders and promote energy conservation and rational energy use throughout the economy. We need to improve the pricing mechanism for energy products and gradually integrate it into the international energy market. We also need to improve the system of compensation for the use of resources as well as the compensation mechanism for degrading the environment. We need to acknowledge the rights and obligations of resource owners and users and protectors of the environment, increase the efficiency of our energy resource use, and offset the damage inflicted on the ecosystem by energy resource development.

5. Implement fiscal policies favorable to sustainable energy development

Economic approaches such as fiscal policies are instrumental in guiding and supporting the implementation of energy strategies and plans. They contribute to promoting energy conservation by widening the application of new energy technologies and products and accelerating the development and utilization of new and renewable energy sources. A resource tax, for example, is key to regulating developers and producers and promoting sustainable regional development; therefore, the tax rate should continue to rise by a reasonable amount. In Japan and many European countries, sales taxes on fuel, energy, and so on have long been levied to promote energy efficiency and technological development, yielding conspicuous energy savings. Some countries even tax CO_2 emissions, and we need to study and draw lessons from these policies.

6. Improve the system and mechanisms for energy management

Optimizing the energy management system and mechanisms should strengthen and improve the macro-control of energy. As economic globalization and the marketization of our energy sector continue, the government's energy management functions need further clarification. We need to improve our ability to make energy plans and coordinate energy policies, organize the implementation of our energy development strategy; promote a balance between total energy supply and demand; optimize the energy structure; increase efficiency; and ensure national energy security through macro-control, market supervision, social management, and public services. Proven practices such as energy-saving scheduling of power generation, electric power demand management, energy efficiency labeling, and energy-conserving government procurement should be studied, improved, and actively promoted to refine mechanisms favorable to energy conservation and efficiency.

7. Incorporate energy development and management into the legal system

The use of legal measures to standardize and regulate energy development and utilization is common practice in many countries. Since adoption of the reform and opening up of policy, the country has enacted the Coal Law, Electric Power Law, Energy Conservation Law, and Renewable Energy Law, as well as a series of supplementary regulations—all of which have achieved salutary results in practice. On this basis, we need to formulate and enact an Energy Law without delay, improve the basic laws in the field of energy as well as special-purpose laws and regulations, and establish a system of energy laws and regulations. We need to strengthen law enforcement in the energy sector and carry out oversight and supervision in accordance with the law in order to guarantee energy development.

8. Further enhance international cooperation on energy resources

As we strive to increase our domestic energy supply, we need to coordinate the domestic development of energy and the opening of our energy industry to the outside world, further strengthen international cooperation on energy resources, combine our importing and exporting operations more effectively, and maximize both international and domestic markets and resources. In our exchanges with other countries relating to energy, resources, and the environment, we need to pursue peace, development, and cooperation; adopt an open, win-win strategy; adhere to the principles of equality and mutual benefit; vigorously work for dialogue and coordination on international energy and environmental policies; promote communication between energy-producing countries and energy-consuming countries; expand cooperation on energy trade and investment; increase exchanges in the areas of energy technology, management, and expertise; and foster mutually complementary exchanges with other countries on resources, secure energy supplies, and common economic development.

ACKNOWLEDGMENTS

I wish to thank Comrades Ma Fucai, Ning Jizhe, Zhou Dadi, and Han Wenke for their assistance in writing this article.

References

[1] Hua Z. *Energy Economics*. Chin. ed. Dongying, Shandong: China University of Petroleum Press; 1991.
[2] National Bureau of Statistics of the People's Republic of China. *China Statistical Yearbook* 2001–07. Chin. ed. Beijing: China Statistics Press; 2001–07.
[3] United Nations Intergovernmental Panel on Climate Change. Fourth Assessment Report: Climate Change 2007.
[4] Hansen J, Sato M, Ruedy R, et al. *Global Temperature Change*. PNAS 2006; 103(14):288–14, 293.

[5] National Bureau of Statistics of the People's Republic of China. *China Statistical Yearbook* 2007. Chin. ed. Beijing: China Statistics Press; 2007.

[6] Building Energy Conservation Research Center of Tsinghua University. *Research Report on Development of Building Energy Conservation in China 2007*. Chin. ed. Beijing: China Architecture & Building Press; 2007.

[7] Yao Q, Chen C. *Clean Coal Technology*. Chin. ed. Beijing: Chemical Industry Press; 2005.

[8] National Development and Reform Commission of the People's Republic of China. Long- and Medium-Term Development Program of Renewable Energy Resources. [EB/OL], 2007-09-28 [2008-02-28], http://www.ndrc.gov.cn/fzgh/ghwb/115zxgh/P020070930491947302047.pdf

[9] Ma X. *Development and Application of Nuclear Energy*. Chin. ed. Beijing: Chemical Industry Press; 2005.

[10] Li J. *Wind Power 12 in China*. Chin. ed. Beijing: Chemical Industry Press; 2005.

Energy Development Trends and Major Energy Conservation Measures*

Weng Shilie, president of Shanghai Jiao Tong University, presenting Jiang Zemin with a school badge at the ceremony at which Jiang was appointed professor in March 1989.

*Article published in the *Journal of Shanghai Jiao Tong University*, No. 3, 1989, a revision of the presentation Jiang Zemin made at the university in March 1989 on his appointment as professor there.

Energy is not only an essential material base for a national economy but also a fundamental condition for human existence; therefore, energy issues have always been important global issues. The Party and government have paid great attention to energy issues, particularly since the Third Plenary Session of the Eleventh Central Committee of the Communist Party of China (CPC), and they have designated energy development as a strategic focus of economic and social development and formulated appropriate principles and policies, thereby significantly developing our country's energy work. However, the rapid development of the national economy and the constant rise in people's living standards have increased energy demand, exacerbating the imbalance between supply and demand, which has become a major restraint on national economic development. Energy shortages now pose a severe challenge to China's modernization drive, and energy issues have become a widespread concern for the whole country.

I. THE ENERGY SITUATION AND DEVELOPMENT TRENDS

1. History of world energy development

World energy use has experienced three major transitions. The first took place in the mid-eighteenth century after James Watt invented the steam engine, which replaced human and animal power with steam power and thus initiated the Industrial Revolution. At this time coal also began to replace firewood as the primary fuel of choice. The second transition began in the 1870s, when electricity gradually supplanted steam as the major power source, leading to capitalist industrialization on a hitherto unprecedented scale. The third transition began in the 1950s, when the large-scale exploitation of cheap oil and natural gas led to the displacement of coal as the most consumed form of energy in the world, thus ushering in the Golden Age of the Western economies in the 1960s. Between 1950 and 1973 world energy consumption tripled, and the proportion of coal in the consumption structure dropped from 60 percent to 32 percent, while that of oil and gas increased from 36.6 percent to 66 percent. Throughout the 1960s, the GNP of the major industrialized countries grew at an average rate of 4 percent to 5 percent per year, and Japan's GNP attained a growth rate of 9 percent to 11 percent. Following the development of atomic energy in the 1940s, the first nuclear reactor went into operation in the Soviet Union in 1954, marking nuclear power's entry into the ranks of the world's energy sources and heralding further changes to the global energy structure. Each time humans have taken a great step forward in energy use and expanded its application, the changes have been accompanied by scientific and technological progress and have provided an extremely powerful impetus to the productive forces, even to the extent of revolutionizing the mode of production.

However, the human race has also encountered three energy crises in the history of energy use. The first occurred in the sixteenth and seventeenth centuries. Due to the rapid development of industrial production in Western capitalist countries, fuel consumption increased, resulting in a severe shortage of firewood in Britain, the Netherlands, and elsewhere that drove the forest resources of these countries to the brink of exhaustion; however, exploitation of coal resources later resolved this crisis. The second crisis occurred after World War I. Due to the war's devastation, coal production plummeted just when every country was trying to develop its national economy, leading to a huge disparity between energy supply and demand. The third crisis occurred in the early 1970s, specifically 1973, when the oil embargo imposed by OPEC in response to war in the Middle East (the Yom Kippur War) caused a worldwide energy shortage. The political background of that energy shortage was the plundering of Third World oil resources by Western industrialized countries, which ignited opposition by the oil-producing countries. The shortage had a very severe impact on Western countries, including the United States, Japan, and some European countries, which suffered serious economic recessions as a result. As these examples demonstrate, energy issues are not only important in constraining economic development but are also key factors influencing the contemporary world political situation.

2. The world energy situation and its development trends

At present, fossil fuel reserves—especially oil, natural gas, and other high-quality energy resources—are gradually being depleted, while major breakthroughs have not yet taken place in the development and use of alternative energy sources. This situation has put us in an energy trough, with a supply shortage at a time when world energy demand is increasing steadily. This global energy shortage is unlikely to be resolved in the short term.

Table 1   Estimated ultimate recoverable reserves of various fossil fuels

Fossil fuel	Type of cost[1]	Ultimate recoverable reserves (TW·a)
Coal[2]	1	560
	2	1019
Oil	1	264
	2	200
	3	373
Natural gas	1	267
	2	141
	3	130
Uranium	1	57
	2	362

It is estimated that the ultimate recoverable reserves of all fossil fuels in the world total 3373 TW·a (1 TW·a equals 1 billion tons of raw coal) (see Table 1); with world energy consumption increasing at a rate of 5 percent per annum, these reserves are expected to run out in one or two centuries (see Figure 1).

In order to address the energy shortage, research and development is currently being undertaken around the world in the following areas:

---

[1]Recoverable resources of a given type are those with an estimated cost equal to or lower than a certain cost (in constant 1975 U.S. dollars).

Coal: Type 1 is $25 per TCE; Type 2 is $25 to $50 per TCE.

Oil and natural gas: Type 1 is $12 per BOE, Type 2 is $12 to $20 per BOE, and Type 3 is $25 to $50 per BOE.

Uranium: Type 1 is $80 per kilogram; Type 2 is $80 to $130 per kilogram.

[2]These figures represent only about 15 percent of the ultimate reserves, because they are already huge for the next 50 years and there are many uncertainties in estimating future coal resources and production technologies.

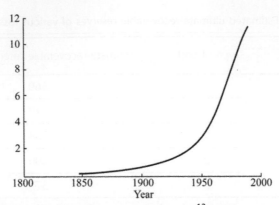

Figure 1   Increase in world energy consumption (10^{13} kW·h)

a. Substituting coal for oil

According to the analyses of international energy experts, oil production will gradually decline after the 1990s due to the depletion of oil reserves. In 2030 global coal production will be double its present level and the energy structure will change, with coal increasing to 29 percent to 34 percent, oil dropping to 19 percent to 22 percent, natural gas making up 15 percent to 17 percent, nuclear energy 23 percent, hydro and geothermal energy 4 percent to 7 percent, and solar energy and other renewable energy sources about 4 percent. However, to protect the environment and use coal more efficiently, 56 percent of coal will be converted to liquid fuel, accounting for approximately 40 percent of all liquid fuel in the future.

Through a process of coal liquefaction, achieved by adding hydrogen to coal under high pressure and at a high temperature, coal can be transformed into 40 percent liquefied oil, 15 percent high-calorific-value coal gas, and solid fuel; this last is strongly bonded and low in ash and sulfur, and can be used as a binder in coal blending for coking and producing coke, thereby opening up broad prospects for the comprehensive use of coal. Sasol's coal-to-liquid plants in South Africa have been in commercial production for many years, and it is estimated that they will reach a production of 30,000 tons per day by the end of this century.

In addition, coal may be made into a coal-oil mixture or coal-water mixture to replace petroleum. This type of technology is developing rapidly, with General Motors in the United States currently conducting research and experiments using 3 μm coal powder as automotive fuel.

b. Developing nuclear energy

One gram of nuclear fuel contains three million times as much power as one gram of coal. One kilogram of U^{235} is equivalent to 2800 TCE, or 2000 tons of petroleum, and it is much easier to store and transport because of its small size and light weight. Furthermore, as the cost of nuclear power is low (the ratio between the cost of coal-generated and nuclear power is 1.25 to 1.70 in foreign countries) and the investment in nuclear power facilities is basically on a par with that of thermal power facilities (including present investment in desulfurization facilities in foreign thermal power plants), nuclear power is developing rapidly. According to statistics, there were 376 nuclear power plants in operation worldwide in 1985. If the more than 500 nuclear power plants under construction were added in, their aggregate power output would be 276.973 GW.

Nuclear power has already become the leading source of power in some countries. For example, it accounts for 64.8 percent of total power capacity in France, 57.8 percent in Belgium, and 42.3 percent in Sweden. The French government has even announced that in the future only nuclear power plants will be developed and no thermal power plants will be built. It is estimated that by 2000, 30 percent of the world's total electricity will be generated by nuclear power, and installed capacity will reach 700 GW. By 2030, nuclear power will account for 23 percent of total world energy resources, and it is likely that the total global installed capacity will reach 2 TW.

The main concern of the general public concerning developing nuclear power is safety. Nevertheless, an analysis in "WASH-1400" indicates that risks posed by nuclear accidents are far lower than those posed by non-nuclear accidents. In terms of the harmful effect of radioactive substances on local residents, a 1 GW nuclear power plant's radiation level of gas and radioactive trace elements is 13 μSv/a, which is one-third of the 47.5 μSv/a radiation level of radium and thorium in the extractor of a coal-burning power plant with equivalent capacity.

However, the prospects for electricity generated by nuclear fission are not great, given that only 7 percent of natural uranium is U^{235} and the rest is U^{238}. This means there are far fewer economically viable uranium reserves than coal reserves, and the future utilization of nuclear energy will rely principally on the following two methods:

(i) *Breeder reactors will be used.* In a breeder reactor U^{238} is used to breed Pu^{239}, Th^{232} is transmutated to become U^{233}, and then U^{233} and Pu^{239} in turn generate heat by means of fission. Fast breeder reactors, which are now being developed, may improve the utilization ratio of uranium to 50 percent to 60 percent. After more than three decades of research, this low-polluting technology is basically mature and becoming economically competitive. According to 1982 statistics, six fast breeder reactor power plants were then under construction worldwide, with a combined production of 5039 MW, while four were in operation, with a total production of 1470 MW. According to the latest statistics, there are now 19 fast breeder reactor power plants in the world, 12 of which are in operation, including the Phoenix reactor in France, which has been running for more than 10 years.

(ii) *Nuclear fusion is the future of nuclear power.* It relies on accelerating nucleons to a very high speed by heating them, using 150 million kW•h, which is a difficult process. However, if we can learn how to control the nuclear fusion reaction and use the deuterium in seawater as fuel, we will have an inexhaustible energy supply in the future. There are 45 trillion tons of deuterium in seawater suitable for use in nuclear fusion, and a 1 GW nuclear fusion power plant consumes only 304 kilograms of deuterium a year. This means the deuterium in seawater is sufficient for human use for over 100 million years.

It was reported recently that British and American scientists are conducting nuclear fusion experiments with deuterium-rich heavy water in test tubes in an attempt to make nuclear batteries, which could be the starting point of human use of nuclear fusion energy. However, it is unlikely that there will be a breakthrough in mastering the technology for large-scale controllable nuclear fusion in the near future. Nevertheless, the ultimate solutions to global energy shortages will probably lie in nuclear fusion and solar energy.

China has just begun to use nuclear energy, and presently two 900 MW generators in the Dayawan Nuclear Power Plant, Guangdong, and a 300 MW generator in the Qinshan Nuclear Power Plant, Zhejiang, are under construction. We therefore expect to achieve considerable development of nuclear energy by 2000. East China, notorious for its severe lack of energy, is the most appropriate area for the development of nuclear power, as it will ease transportation pressures there.

c. Utilizing solar energy

Solar energy is the cleanest energy resource. Five ten-billionths of the energy produced by the sun, approximately 173 PW, travels to the earth in the form of electromagnetic waves; 30 percent of this energy is reflected directly back into space. The rest penetrates the atmospheric layer and continues on toward the earth's surface, where 47 percent is absorbed by the atmosphere and the earth's surface, warms them, and then returns to space in the form of long waves. The remainder becomes the driving force for wind, waves, tide, and water circulation, ultimately radiating back into space. Only 0.023 percent of solar energy, or 40 TW, enters the biological system through photosynthesis, and only 0.63 percent of this is consumed by human beings.

The biggest problem humankind faces in using solar energy is twofold. First, the density of solar energy is only 200 W/m^2, and second, it is affected by climate. Solar energy is used mainly in the following forms: light $\rightarrow$ heat utilization, light $\rightarrow$ electricity utilization, and light $\rightarrow$ chemical utilization.

At present there are nine solar power plants in the world, with a total capacity of 15.8 MW, the largest of which is in the United States, in Barstow, California, with a capacity of 10 MW; the Soviet Union, however, is in the process of building a 100 MW solar power plant. It is also reported that the United States is planning to launch a solar power satellite carrying a PV cell with an area of 4.8 km × 9.6 km and electricity generation capacity of 500 MW in 2000, and that it will launch 100 such satellites by 2025. This technology is on the edge of a breakthrough and could become a major channel for the utilization of solar energy. In 1984 a 4 MW PV power station generating electricity with PV cells was built in the United States and successfully integrated into the power grid. Japan is investing massively in research of PV cells and plans to generate 300 MW of power, worth a total of $3 billion, by 1995.

The region most endowed with solar energy in our country is Tibet, followed by Gansu and then Beijing. By the end of 1985 China had already produced solar-powered water heaters with a combined surface area of 500,000 m^2, built 231 solar-powered buildings with a total floor space of 82,381 m^2, and put 100,000 solar-powered cooking stoves into use. According to estimates, solar-powered water heaters in China will have a total area of 88 million m^2 by 2000. A 10 kW PV power station has been built in Beijing in cooperation with West Germany, while another has been built in

Lanzhou in cooperation with Japan. The Xiangtan Electrical Manufacturing Factory built an experimental solar power plant with a capacity of 2 kW in cooperation with the United States in 1984.

d. Utilizing biomass energy

According to estimates, the total energy stored in the world's plants each year, if converted to electricity, would be equal to 500 MW•h/a per capita, which is forty times more than present per capita annual energy consumption; this form of energy therefore clearly possesses great potential.

Many countries in the world are currently producing ethanol as an oil substitute, primarily from plants. According to estimates, Brazil produced more than one billion liters of ethanol fuel in 1986, equivalent to about one-third of its oil consumption.

At present China's application of biomass energy mainly takes the form of producing methane via anaerobic processing, with five million rural households nationwide using methane, and nearly twenty-five million rural residents using high-grade methane fuel.

e. Developing hydrogen energy

Hydrogen energy, one of the cleanest energy sources, is being developed in advanced industrialized countries around the world as a substitute for oil and other fuels. Currently, the main method of producing hydrogen is by cracking natural gas and residual oil. However, in the 1990s, coal gasification and new electrolysis methods will be developed to produce hydrogen. The coming trend is to use high-temperature reactors to produce hydrogen from lignite or use coal gasification and nuclear-powered electrolysis technology. Hydrogen accounted for 0.4 percent to 0.5 percent of secondary energy sources in the 1980s and is expected to reach 4 percent to 6 percent by 2005.

f. Exploiting and using geothermal energy, wind energy, and various forms of ocean energy—such as tidal energy, wave energy, energy produced by temperature differentials, and salinity gradient energy

By 1984 the total installed capacity of all the geothermal power generators in the world had reached 3.4 GW. The largest such power plant is the Geysers Geothermal Power Plant in the United States, with a total installed capacity of 1.3 GW, while the Yangbajing Geothermal Power Plant in Tibet has an installed capacity of 10 MW.

The Rance Tidal Power Plant, built in France in 1968, is the largest of its kind in the world and is equipped with 24 10 MW reversible bulb–type generators. China has a 600 kW experimental tidal power plant in Zhejiang.

In brief, here are some current world energy development trends:

(i) Energy production will not increase significantly in the near future. It is predicted that oil production will reach a peak in the 1990s, after which the decline of world oil production could possibly cause an energy crisis.

(ii) Energy consumption will continue to rise. World energy consumption was 10.6 billion TCE in 1985, and demand is expected to reach 25 billion TCE in 2000; therefore, unless energy consumption is controlled, acute energy shortages will emerge.

(iii) The energy structure will undergo great changes, with a resurgent, coal-dominated structure replacing the current oil-dominated one.

(iv) Great effort will go into developing new types of energy, particularly nuclear energy, whose share in the energy structure will gradually increase.

(v) To alleviate the contradictions in energy supply and demand and protect the environment, energy conservation will attract every country's total attention. At present, energy conservation is viewed worldwide as a fifth energy source and is being actively developed alongside the use of coal, oil and natural gas, hydropower, and nuclear power sources.

3. The condition and characteristics of our country's energy resources

China is rich in energy resources and ranks third in the world in total geological energy reserves. However, it is an energy-poor country as far as per capita reserves are concerned, with only half of the world average, one-tenth as much as that of the United States, and one-seventh that of the Soviet Union. With regard to primary energy reserves, China is rich in coal but poor in oil. The average global ratio of recoverable reserves of coal to those of oil and natural gas is approximately 4 to 1, but this ratio is much larger in China, which means that low-quality energy predominates here.

a. Reserves

Coal: At 1.44 trillion tons China's geological coal reserves rank third in the world. By the end of 1983 China's proven coal reserves were 770 billion tons, also ranking third in the world. Of the 175 billion tons of recoverable coal reserves, 100 billion tons are being exploited. China's current annual prospecting capability is 5 billion tons.

Oil and natural gas: China's geological oil reserves are 60 billion tons, but the proven reserves are only slightly more than 7 billion tons, while the proven recoverable reserves stand at only 2.33 billion tons. China's proven reserves of natural gas are approximately 500 billion m^3. The present annual oil-prospecting capability is only around 150 million tons.

Hydropower resources: China has the largest hydropower reserves in the world, approximately 680 GW, accounting for one-third of the world's total. Only around 5 percent of the 378 GW of recoverable reserves have been exploited. According to 1985 statistics, the total installed capacity of national hydropower amounted to 26.41 GW. This capacity accounted for 30.3 percent of the combined installed capacity of hydropower and thermal power plants and generated 92.4 billion kW•h, accounting for 22.5 percent of total power generation. It is expected that the installed capacity of our country's hydropower plants will reach 80 GW by 2000 and that annual power generation may reach 250 billion kW•h.

Our country plans to build ten hydropower plants on the middle and upper reaches of the Yellow River; in the Hongshui River basin; on the Jinsha, Yalong, Dadu, and Wujiang rivers; on the upper reaches of the Yangtze River; on the Lancang River; and in western Hunan, Fujian, Zhejiang, and Jiangxi provinces. These plants will have a total installed capacity of 187.52 to 193.32 GW and will generate 973.3 billion to 998.8 billion kW•h of electricity annually, significantly relieving China's electricity shortage.

b. Distribution of energy resources

(i) Coal is concentrated in north China, with more than 60 percent of the national coal reserves found there; Shanxi alone possesses about one-third. Areas where industry is concentrated—east and south central China—contain less than 10 percent of the country's coal reserves.

(ii) Oil is concentrated mainly in northeast China, which has more than 50 percent of national reserves.

(iii) Natural gas is distributed largely in Sichuan and Guizhou, where around two-thirds of national reserves are found.

(iv) Hydro resources are distributed mainly in southwest China, accounting for over 70 percent of national reserves.

Eight provinces and one municipality directly under the central government in south China—namely, Jiangsu, Zhejiang, Jiangxi, Hunan, Hubei, Guangdong, Guangxi, Fujian, and Shanghai—though renowned for being poor in energy resources, are the areas that consume the most energy.

Because of the vast distances between energy sources and energy-consuming areas, the transportation volume of energy resources has been increasing. According to statistics, in 1984 the transportation of energy resources accounted for approximately 43 percent of total railway traffic and 47 percent of total waterway traffic. However, a considerable degree of loss occurs during transportation, with the average loss rate of coal being 4 percent to 5 percent during railway transportation and 6 percent to 7 percent during waterway transportation. Therefore, more investment is needed in railways, port construction, and shipbuilding to facilitate the transportation of energy resources.

c. China's energy production situation

China's energy production statistics are shown in Table 2.

The data in Table 2 show that the growth rates of oil and natural gas production have fallen steadily.

d. China's total energy consumption and energy consumption per 10,000 yuan of output value

Table 3 shows China's total energy consumption and consumption per 10,000 yuan of output value.

Table 3 shows that the energy consumption per 10,000 yuan of output value fluctuated, mainly due to the effects of changes in factors such as the industrial structure, the product mix, and GNP.

e. Characteristics of China's energy consumption

(i) *Dominance of coal.* According to data, coal accounted for 73.85 percent, oil 17.09 percent, natural gas 2.25 percent, and hydropower 4.81 percent of the 764 million TCE of national energy consumption in 1985. However, a coal-dominated energy structure is associated with low energy efficiency and serious environmental pollution.

Table 2   China's energy production statistics for 1949, 1979, and 1988

Year	Coal (in millions of tons)	Oil (in millions of tons)	Natural gas (in billions of m³)	Electricity generation (in billions of kW·h)
1949	34.23	0.121	0.007	4.31 (7)
1979	635	106	14.51	281.95 (501)
1988	960	137	13.9	539
Annual growth rate (1949–79) (%)	10.43	25.34	125.70	14.95
Annual growth rate (1979–88) (%)	4.6	2.88	-0.30	7.45

Table 3   China's total energy consumption and energy consumption per 10,000 yuan of output value

Year	Total energy consumption (in millions of TCE)	Energy consumption per 10,000 yuan of output value (TCE)
1950	32	6.3
1955	70	7.18
1960	302	14.8
1965	189	9.46
1970	293	9.45
1975	454	10.08
1980	603	9.11
1985	764	6.46

(ii) *Dominance of industry in energy consumption.* Statistical data for 1980 show that industry consumed a greater portion of energy in China than it did in developed countries (see Table 4).

(iii) *Large proportion of noncommercial energy.* Biomass energy is the most important energy source in rural areas. In 1980 the production of biomass energy such as straw in rural areas was about 255 million TCE (this portion of energy resources is not included in national statistics). If these noncommercial energy resources are factored in, the total energy for agricultural production and household use in rural areas would account for 38.4 percent of national energy consumption.

(iv) High energy consumption and low energy efficiency of electrical appliances. According to 1979 data, the energy efficiency of China's electrical appliances and equipment was low compared with that of developed countries. The overall energy efficiency was 57 percent in Japan, 51 percent in the United States, and 30 percent in China. Energy efficiency is a reflection of a country's technological level.

(v) Low per capita energy consumption. Per capita energy consumption is a comprehensive indicator of economic development and living standards in every country. Table 5 shows the per capita energy consumption of the world's major countries in 1980.

Table 4 Comparison of China's energy consumption with that of developed countries (%)

Country	Industry	Transportation	Civil use	Energy authority allocation	Others
China	62.9	3.8	25.2*	7.0	1.1
United States	27.5	31.6	29.9	6.2	4.8
Japan	50.4	19.5	21.0	5.6	3.5
W. Germany	35.1	19.6	37.5	5.6	2.5
Britain	30.4	23.1	38.4	5.7	2.7
France	35.0	21.6	33.4	5.7	3.3

*Including agriculture 7%.

Table 5 Major countries' per capita energy consumption in 1980 (TCE/annum)

World average	United States	W. Germany	Soviet Union	Britain	France	Japan	Italy	China	India
1.955	10.410	5.727	5.595	4.835	4.351	3.690	3.318	0.596	0.191

Table 5 shows that our per capita energy consumption is just 30 percent of the world average.

According to a study by the U.S. Agency for International Development, China will reach the living standards of a moderately prosperous society in 2000, with an annual per capita energy consumption of approximately 1.2 to 1.4 TCE. Electric power in particular will have a greater influence on people's living standards. Based on this forecast, China's energy production will need to reach 1.44 to 1.68 billion TCE by 2000.

The Twelfth National Party Congress announced that China would do its utmost to quadruple its gross industrial and agricultural output value in the 20 years between 1980 and 2000 while continually improving economic efficiency. However, during the same period, it is expected that energy production will only double, due to the significant consumption of investment, labor, materials, equipment, and energy required for the exploitation, processing, transformation, and transportation of energy resources. The energy industry provides fuel and power to energy-consuming sectors, which in turn provide materials and equipment to the energy industry. The two sides are thus in a symbiotic but mutually restrictive relationship. Because coal is the predominant energy source in China and energy reserves are located far from the energy-consuming cities, more investment is needed in energy transportation. In the Soviet Union, for example, investment in the energy system—including exploitation, processing, transformation, and transportation—accounted for 28 to 34 percent of total industrial investment in the 1970s, while labor input in the energy system accounted for 19 percent to 22 percent of total labor input. In the early 1980s the exploitation and transportation of oil and natural gas alone required an average of 7.5 million tons of pig iron, 12.5 million tons of steel, 10 million tons of steel materials, 3.5 million tons of large-caliber steel pipes, and 15 million tons of cement per year. In China during the Sixth Five-Year Plan period (1981–85), the exploitation of 1 TCE of energy cost 375 yuan (excluding processing, transformation, and transportation). Therefore, the shortage of funds, materials, and equipment has become a very important constraint on the growth of energy production.

If gross industrial and agricultural output value is to quadruple while energy production can only double, the only solution is energy conservation. Therefore, the general principle of China's energy policy is to place equal emphasis on exploitation and conservation. However, maintaining the energy elasticity coefficient at 0.5 for 20 years is very uncommon in the world. To attain this goal, the annual average energy conservation rate must reach 3.7 percent; but the rate in most countries is only 2 percent. Furthermore, China used to rely primarily on indirect energy conservation such as economic restructuring. During the Sixth Five-Year Plan period, direct energy conservation by companies that reduced unit energy consumption accounted for only 30 percent of total energy saved. In the future, therefore, we need to rely more and more on technological progress and scientific management, genuine ability, and hard work, as well as greater investment. We also need to make the Chinese people realize that we cannot solve the problem of insufficient energy supplies quickly, that this problem may at times be very serious, and that we cannot afford to let our guard down if the situation improves temporarily.

At present, Shanghai and China as a whole are experiencing a severe shortage of energy supplies because of economic overheating and soaring household energy demand in recent years (for example, electricity consumption by households has grown at more than 10 percent annually, and even 30 percent in some places). However, we must not overlook the effects of inadequate energy conservation efforts in recent years. As we improve the economic environment, rectify the economic order, and solidify reform, the current strained energy situation will gradually improve, but it will be difficult to eliminate energy shortages anytime soon. Therefore, we must be mentally prepared for a long-term energy conservation drive.

II. CHINA'S ENERGY CONSERVATION POTENTIAL

1. Forecasts for China's energy conservation potential

The above section states that energy conservation in China is a very difficult task. Is it possible for us to accomplish this task? This possibility is analyzed below.

A forecast of energy conservation potential is often conducted using the elasticity coefficient method and the sector analysis method.

a. Predicting energy conservation potential using the energy elasticity coefficient method

The energy elasticity coefficients for China from 1978 to 1985 are presented in Table 6.

From 1978 to 1985 China's average energy elasticity coefficient was around 0.4. In 1988 its energy consumption per 10,000 yuan of industrial output value decreased by 5.81 percent, and this figure reached 8 percent in Shanghai. On this basis it should be possible to maintain the energy elasticity coefficient at 0.5 from 1989 to 2000.

Table 7 shows the energy elasticity coefficients of several industrialized developed countries:

Table 6   China's energy elasticity coefficient statistics for 1978 to 1985

Year	1978	1979	1980	1981	1982	1983	1984	1985
Energy elasticity coefficient	0.74	0.29	0.40	–0.21	0.43	0.55	0.52	0.49

Table 7   Energy elasticity coefficients of major industrialized developed countries

Country	Year				
	1950–60	1960–70	1970–73	1973–77	1977–80
United States	0.75	1.15	0.98	0.40	–0.13
Japan	0.99	1.13	0.82	0.30	0.54
West Germany	0.65	0.99	0.91	–0.23	0.40
Britain	0.48	0.58	0.48	–5.11	–1.38
France	0.80	1.02	0.96	–0.26	0.21

The energy elasticity coefficients in Table 7 may be divided into three levels:

(i)   Energy elasticity coefficient lower than 0.6. This situation, which prevailed in Britain in the 23 years from 1950 to 1973 and in the other countries after the energy crisis in 1973, is characterized by slow economic growth. The annual growth rate in Britain was 2.6 percent, and that of the other countries during this period was only 2 percent to 4 percent.

(ii)   Energy elasticity coefficient at the intermediate level of 0.65 to 0.85. The economic growth rate for this level was 5.4 percent to 8.7 percent, roughly equivalent to the planned growth rate of China from 1980 to 2000.

(iii) High energy elasticity coefficient, between 0.9 and 1.1. At this level the speed of economic development is more rapid, with more dependence on energy imports.

Based on the above analysis, it is advisable for China to maintain the energy elasticity coefficient at an intermediate level, namely, 0.7 to 0.75. From this perspective, if China's energy production doubles by 2000, reaching 1.2 billion TCE, the energy supply will fall short of the required quantity of 1.6 to 1.7 billion TCE. Therefore, China's energy conservation task is formidable.

b. Predicting energy conservation potential using the sector analysis method

Because most of China's industrial technologies and equipment are outdated and business management skills generally underdeveloped, energy consumption is high and generates substantial waste. According to data from the World Resources Institute on energy consumption per U.S. dollar of GNP among the world's top ten energy-consuming countries (excluding the Soviet Union) in 1987, China's consumption was 4.98 times greater than France's and 1.64 times more than India's. Although there are some reasons why a direct comparison is not appropriate, this analysis nevertheless reveals China's energy conservation potential. If China can surpass India's level of energy consumption just a little, it could quadruple its 1980 GNP with current energy resources. In addition, there is a huge gap between China and the developed countries in terms of energy efficiency between different sectors (see Table 8). Table 9 compares the energy efficiency of equipment and appliances.

Table 8 Energy efficiency of different sectors in China, Japan, and the United States (%)

Country	Electricity generation and transmission	Industry	Transportation	Household use	Overall efficiency
Japan	30.0	76.0	22.4	75.4	57
United States	30.6	75.1	25.1	75.1	51
China	23.9	35.0	15.2	25.5	30

Table 9 Energy efficiency of equipment and appliances in China and developed countries (%)

Item	Developed countries	China (1979)
Efficiency of thermal power generation	35–40	28 (including small generators)
Thermal efficiency of industrial furnaces	70–80	55
General efficiency of the iron and steel industry	50–60	28
Thermal efficiency of synthetic ammonia production	50–60	25
Thermal efficiency of railway transportation	20–25	6–8
Thermal efficiency of cooking devices	60–70	15–18 (urban coal stoves) 10 (rural firewood stoves)

The experience of some developed countries demonstrates that the reduction in unit energy consumption for some major products envisaged by China is practicable. Take coal consumption in supplying power, for example. This rate could decrease from 448 g/kW•h in 1980 to 352 g/kW•h in 2000, that is, a reduction of 96 g/kW•h in 20 years. In the Soviet Union, the rate decreased by 118 g/kW•h in the 17 years from 1962 to 1979. Energy consumption per ton of steel could decrease from 1.3 tons in 1980 to 0.95 tons in 2000, a reduction of 0.35 tons in 20 years. In Britain, this rate decreased by 0.4 tons in the 20 years from 1960 to 1980. Moreover, the energy consumption levels of industrial products in some areas of China are already on a par with or close to world advanced levels, which shows the potential for energy conservation. In Shanghai for example, the energy consumption in producing 75 percent ferro-silicon at the Shanghai Ferroalloy Factory is 8124 kW•h/t, compared with the average level of 8400 kW•h/t in industrialized developed countries. The Shanghai Solvents Factory's total energy consumption in producing solvents is 1.376 TCE/t, in comparison with the average level of 1.607 TCE/t in industrialized developed countries.

Energy consumption levels vary greatly in different areas of the country due to differences in industrial structure, product mix, technology, and managerial expertise. In 1988 the energy consumption per 10,000 yuan of gross industrial output value was 2.21 TCE in Shanghai, 2.56 TCE in Zhejiang, 11.58 TCE in Ningxia, and 10.91 TCE in Shanxi, with the national average being 4.86 TCE.

In conclusion, China has huge potential for energy conservation, and the target of quadrupling GNP while only doubling energy supply should be achievable.

2. Basic causes of high energy consumption in China

a. Dominance of coal in the energy structure

The energy structure of the world today is dominated by oil and natural gas. Used as energy and raw materials, in most circumstances coal is much less efficient than high-quality energy resources such as oil and natural gas. For example, the production of one ton of synthetic ammonia requires only 1.14 TCE of natural gas, with an overall efficiency of 69 percent; however, it requires 1.86 to 2.28 TCE of coal, and overall efficiency is only 34.5 to 42.3 percent.

As far as electricity generation is concerned, when burning oil or natural gas, a boiler's efficiency can be as high as 92 percent, but is only 83 percent to 88 percent when coal is used. In addition, with the extra equipment needed to transport and grind coal, and remove dust, the plant's electricity consumption will be 3 percent to 5 percent greater when using coal than when using oil.

b. High proportion of industries with high energy consumption in the industrial structure

In general, the energy consumption per 10,000 yuan of output value in heavy industry is four to five times more than that of light industry. A large proportion of industries in China, such as the metallurgy, chemical engineering, and building materials industries, involve high energy consumption; these industries account for 47.9 percent of total energy consumption in all industries, while their output accounts for only 20.9 percent of the total national industrial output. By contrast, industries with low energy consumption have not been fully developed.

c. Low level of energy utilization technology and outdated equipment

Boilers, fans, water pumps, and other widely used power equipment are good examples. They consume approximately one-third of the total energy consumed by the country, but their actual efficiency is only 40 to 60 percent, compared with an efficiency of 75 percent to 85 percent in foreign countries. This is a huge difference.

d. Outdated production processes

Japan's iron and steel industry now uses large, pressurized electric furnaces in processes such as converter steelmaking; continuous ingot casting; and large-scale, continuous automated steel rolling. Using this technology, coupled with high-quality raw materials and fuels, the country's average energy consumption per ton of steel is below 0.7 TCE, compared with China's rate of 1.8 TCE in 1984.

The production of synthetic ammonia using heavy oil and natural gas as raw materials, and industrial turbines in place of electric motors, requires very low energy consumption and almost no electricity (5 to 20 kW•h of electricity per ton of ammonia); therefore, energy consumption per ton of ammonia is only 1.04 to 1.07 TCE.

e. Poor energy management

(i) Energy is allocated irrationally. If oil rather than coal were used as raw materials for making chemicals, energy efficiency would increase by 40 percent to 80 percent. However, according to 1984 Chinese statistics, about 30 percent of total oil production was used as fuel in industrial furnaces or power generation boilers, and only about 5 percent was used for making chemicals. Furthermore, natural gas and liquefied petroleum gas should be supplied to urban households as substitutes for coal. Thus, more work remains to be done in this area.

(ii) Energy supply channels are not stable; therefore, they are unable to supply energy of the required quality and in the designated quantities. Low-quality coal is still being transported over long distances, which causes production difficulties for factories and puts pressure on the transportation system. In other countries coal is generally transformed into electricity for transmission at power plants near the mines.

(iii) Enterprises are responsible for losing considerable power. Leakage, unstable production, breakdowns, and idle running of equipment waste a lot of energy.

(iv) Production layout needs to be further optimized so that energy production bases are close to major energy-consuming enterprises, and steel plants and iron plants should be placed together.

The above problems are the main causes of high energy consumption in China.

III. MAJOR ENERGY CONSERVATION MEASURES

1. Increase energy conservation publicity and raise public awareness

The key to energy conservation is advocating the spirit of saving every kW•h of electricity, every ounce of coal, and every gram of oil.

a. First of all, we must publicize the importance of energy conservation.

Energy conservation can boost production. According to estimates, China could increase production by 20 percent to 30 percent using existing equipment if energy resources and raw materials were sufficient.

According to international statistical data, one kW•h of electricity saved can generate output value 44 times its own value.

Energy conservation is beneficial for accumulating capital. In China's industrial production, the cost of energy accounts for 8 percent to 9 percent of the total output value. In the iron and steel and the nitrogenous fertilizer industries, energy accounts for 30 percent to 70 percent of cost. Therefore, energy conservation may reduce costs and accumulate capital.

Energy conservation is beneficial for promoting technological progress. The degree of energy utilization is an important indicator of a country's scientific and technological modernization, and energy conservation is a reflection of national technological progress. Energy conservation relies on and promotes scientific and technological progress.

Energy conservation is beneficial for relieving transportation shortages. In China, coal transportation accounts for over 40 percent of total railway freight, and even 70 percent of traffic on certain main routes. Therefore, energy conservation can help reduce traffic levels.

Energy conservation is beneficial for protecting nature and can delay the depletion of fossil fuels and minerals. Less cutting of firewood in rural areas will protect forest resources and prevent soil erosion.

Energy conservation can help protect the environment. Waste gas, water, and solids given off during the production, processing, transformation, and use of energy resources are a major source of pollution. Every year about 15 million tons of SO_2 and dust are emitted into the atmosphere from the burning of coal, while coal-burning power plants release 40 million tons of ash residue as well as carbon monoxide, nitrogen oxide, benzene, hydroxybenzene, cyanogens, bitumen, tar, benzopyrene, and other toxins and carcinogens into the air, causing enormous harm to human beings.

b. We need to publicize the current energy situation as well as the principles, policies, laws, and regulations of the CPC and the government in order to help the public understand that energy conservation is a long-term state policy and make conscious efforts to limit consumption and save energy.
c. We need to spread the knowledge of how to conserve energy, promote new energy-efficient products and technologies, and educate people on how to save energy.

Using good publicity concerning energy conservation, we need to organize people to take part in practical energy conservation activities. The experience of the Changchun-based First Automotive Works (FAW) in energy conservation measures is therefore provided here for reference.

(i) Turning off the lights when leaving a room, using low-wattage bulbs in place of high-wattage ones where possible, and turning street lights on and off with light-control switches in order to save electricity used for lighting. These measures reduced FAW's electricity consumption by about 2 percent.

(ii) Stopping and preventing the leakage of gas, oil, water, and coal powder. Leakage not only wastes large quantities of energy resources but also pollutes the environment. FAW motivated employees to prevent leakage by organizing competitions like "zero leakage teams," "zero leakage workshops," and "zero leakage divisions." They also shut off workshop master valves and machine valves during nonworking hours to eliminate leakage, with remarkable economic benefits. With the above measures, FAW reduced the leakage rate of compressed air from 40 percent to 60 percent to less than 30 percent (compared with 15 percent to 20 percent in the Soviet Union), which is equivalent to a saving in electricity for the whole factory of about 4.6 percent (air compressors consume 23 percent of the total energy consumed by FAW).

(iii) Organizing full-load continuous production, especially for the cupola furnace, thermal treatment furnace, and heating furnace, to prevent extra energy consumption caused by starting up, shutting down, and keeping the furnace empty. Energy consumption of a furnace under full load is less than twice the consumption under half load; therefore, full-load production saves energy. Measures were also taken to prevent equipment from running idle, in order to eliminate energy loss during zero-load operations.

(iv) Operating carefully and reducing waste products, especially cast and forged parts. Producing waste products wastes energy.

(v) Operating carefully and reducing malfunctions, especially the malfunctioning of boilers at power plants, since every startup and shutdown of the boiler consumes more than ten or even several tens of tons of raw coal.

(vi) Repairing thermal equipment and improving the insulation of pipelines. The energy loss of a poorly insulated 40 mm diaphragm valve is

0.208 TCE a year, and that of a 250 mm diaphragm valve is 1.445 TCE a year.

(vii) Organizing competitions on minor energy conservation indicators in the secondary energy production department.

All these are feasible and effective methods, which basically require no further investment, obtain instant results, and can easily be adapted by other enterprises.

2. Rely on scientific and technological progress and vigorously carry out upgrading to conserve energy

a. Enthusiastically making full use of energy and improving energy efficiency

There are many ways of using energy to the full capacity; the main methods used at present are listed below.

(i) *Promoting cogeneration and central heating.* Using high-efficiency boilers for central heating at power plants in place of low-efficiency small boilers that supply heat separately can yield significant energy savings.

The thermal power plant at FAW, which was designed by Soviet experts in the 1950s, is one of the first central heating facilities in China. Currently, this plant supplies heat to buildings with more than 3 million m^2 of floor space within a radius of 7 km, replacing more than 40 small boilers. In the early 1970s the plant took the lead in converting a condensing turbine into one operating in low-vacuum conditions. In winter the steam given off by the steam turbine is used to heat water, which in turn heats the factory dormitory. This measure alone saves 26,000 tons of standard coal a year. With cogeneration, the overall efficiency of the plant rose to 67 percent, 39 percent points higher than the rate of 28 percent in most condensing power plants, bringing remarkable economic benefits.

The Shanghai Gaoqiao Petrochemical Thermal Power Station also experienced remarkable economic benefits by adopting cogeneration and central heating. Its coal consumption in power generation was reduced to 281 g/kW•h in 1981, saving 34,200 tons of standard coal annually.

Central heating not only saves energy but can also significantly reduce environmental pollution, because large power plants have better and more efficient vacuuming equipment, which removes dust more effectively.

(ii) *Using industrial waste heat.* There are many different kinds of industrial waste heat, including waste gas and hot water (condensed water) emitted by energy-consuming equipment, hot smoke, exhaust gas from operations in the chemical and metallurgical industries, heat from chemical reactions, and hot solids. In China much more waste heat could be used. Coal gas in steel factories, exhaust gas from refineries, and various inflammable gases in chemical factories are not fully used through recycling. According to 1980 statistics, 8.2 billion m^3 of blast furnace coal gas was discharged from 36 iron and steel factories, equivalent to the loss of 930,000 TCE, while 645 million m^3 of coke coal gas was discharged, equivalent to the loss of 400,000 TCE.

Statistics show that in Shanghai, of the total 61.76 million GJ/a of industrial waste heat, 28.81 million GJ/a is recycled, leaving 53.3 percent unused.

At FAW, the greasy exhaust vapor from the steam hammer is recovered, cleaned, and then reused by the power plant, raising the thermal efficiency of the forging hammer from approximately 3.5 percent to about 70 percent.

Using industrial waste slag and other by-products is another way to use energy comprehensively. For example, FAW uses the coal tar from the coal gas plant as boiler fuel and then recycles the carbon content in the ash from the boiler by making the ash into bricks.

(iii) *Using steam pressure differentials to generate electricity.* China has adopted a regulation requiring that any boiler operating at and above 10t/h with a stable steam supply load should generate electricity by means of steam pressure differentials. Power generation with steam pressure differentials does not require a lot of investment (only the addition of a back-pressure steam turbine), and the generation of 1 kW•h of electricity consumes only 250 g of standard coal, which is significantly less than the coal consumption of a thermal power plant. At present, Shanghai has established 47 pressure generators, capable of generating 120 million kW•h of electricity a year.

(iv) *Developing full use of stone coal and coal gangue.* If coal gangue is made into cement or bricks, the energy consumed in making 10,000 bricks will drop from 1.5 t to 0.2 t. Coal gangue contains vanadium and other precious metals. If coal gangue is burned in a fluid-bed furnace, the resulting steam may be used to generate electricity and supply heat, and precious metals may be extracted from the slag.

(v) *Gradually promoting the use of coal gas in cities and marsh gas in rural areas.* Changing the energy structure of household cooking stoves can increase energy efficiency by 20 percent to 30 percent and at the same time improve household sanitary conditions and reduce environmental pollution. In addition, the cofermentation of dung and weeds to generate biogas may improve fertilizer efficiency and benefit the soil.

b. Renovating energy-consuming equipment to improve its efficiency

This is a broad area. At present we need to focus on renovating the major energy-consuming equipment that is in wide use and consumes large quantities of energy.

(i) *Renovating inefficient industrial boilers.* China currently has more than 250,000 boilers that consume 300 million tons of coal a year, with an average vaporizing capacity of 2.5 t and an average efficiency of about 50 percent. In addition to adopting central heating, replacing small inefficient boilers with larger efficient boilers, and generating electricity with pressure differential, industries must also upgrade small boilers by such means as improving the grate and combustion chamber structure, renovating the furnace arch, reinforcing furnace wall insulation, and installing heat-pipe heat exchangers in the smoke flue to recycle the heat trapped in the smoke. Experience acquired from the Sixth Five-Year Plan period shows that these measures can increase boiler efficiency by 10 percent to 15 percent.

(ii) *Upgrading industrial kilns.* Industrial kilns in China consume about 100 million tons of coal a year and account for a large share of coal consumption in every industry: 79.6 percent in metallurgy, 87.7 percent in building materials, 53 percent in the chemical industry, 33.81 percent in the nonferrous metals industry, 14.35 percent in machinery, and 15 percent in light industries. Upgrading is oriented toward using advanced combustion systems; high-strength, anti-corrosion, and low heat capacity refractory materials; high–performance insulating materials; and automatic and energy-efficient auxiliary equipment.

(iii) *Upgrading and replacing fans and water pumps.* According to 1982 statistics, electricity consumption of fans and pumps accounted for about 25 percent of the country's total electricity consumption. Most of these fans and water pumps were made in the 1950s or 1960s and have an efficiency of only 50 percent to 60 percent. In contrast, the efficiency of a new type of fans made in China is over 80 percent, and that of a new type of water pumps is over 70 percent. Therefore, upgrading and renovating fans and pumps is a very important aspect of energy conservation.

c. Reforming outdated and inefficient processes and adopting advanced energy-saving processes

This is a very important measure to save energy. When new plants are designed in the future, energy conservation will be an important indicator of technological progress. FAW, an old automotive factory, is currently planning to use steam and hot water (at 150°C and 70°C respectively) to replace steam in the original process. It is cooperating with Tsinghua University and Siemens of the Federal Republic of Germany to build a 200 MW, small, low-temperature nuclear heating reactor (equivalent to two 130 t/h boilers) to heat water. If the plan succeeds, it will not only benefit the environment but also save 150,000 TCE each year. Moreover, FAW will replace outdated equipment such as steam hammers and coal gas heating furnaces, which consume excessive energy, with mechanical forging machines and power frequency heating furnaces. It will replace forgeable cast iron with ductile cast iron and eliminate the annealing process to save more electricity.

Other industries, notably light industries are adopting multi-effect evaporation, liquor raw material fermentation, and low-temperature distillation, along with other technologies; similar work can be done in every industry. All in all, improvements in processing will lead to higher energy efficiency in enterprises.

d. Using and promoting new energy conservation technologies and products

In recent years China has imported and developed many new energy conservation technologies and products, such as the 20 kinds of electricity conservation technologies and products promoted by the government, as well as energy-saving products promoted by the State Machinery Commission. Several such technologies and products in wide use and with good energy savings are listed below:

(i) With regard to new light sources, Shanghai Huguang Light Factory has recently developed a 2D energy-saving lamp. According to the factory, a 16W 2D lamp consumes just 20W of electricity (including 4W consumed by the ballast) but has the luminous flux of an 85W incandescent bulb, saving 65W of electricity per bulb. If the factory turns out one million such bulbs annually and all of them are used in households, 71.17 million kW•h would be saved every year. If half of them were used in factories, mines, and hotels, the estimated saving would be 172.6 million kW•h per year.

Use of combined sodium and mercury lamps can also increase luminance and save 30 percent of electricity.

(ii) Using new far-infrared heating technology, a drying furnace could save 30 percent to 50 percent of electricity. The electrical resistance far-infrared radiators now in use produce the best results.

(iii) Electric motors of all kinds consume around 40 percent of all the electricity generated in China; replacing JO_3 electric motors with Y-series electric motors would improve efficiency by 2.5 percent (the efficiency of Y-series electric motors is 88.263 percent).

(iv) Using new aluminum silicate fiber as a heat insulator could reduce the heat loss of all kinds of furnace walls. Using polycrystalline mullite refractory fiber as furnace lining could also reduce furnace heat loss, and because of its specific low heat capacity, this material is especially effective in furnaces that are frequently turned on and off.

e. Actively developing and importing new energy conservation technologies and products

New breakthroughs in energy utilization technologies often usher in a new era of energy utilization and significant upgrading of industrial products.

To improve the efficiency of power plants, we are in the process of developing combined gas-steam circulation technology, which will raise power plant efficiency to 4 percent, far above that of an ordinary condensing power plant. China is building a 200 MW combined gas-steam circulation power plant that will burn natural gas as fuel in the gas turbine at Shengli Oilfield in cooperation with the Federal Republic of Germany. Some overseas power plants burn coal gas as fuel, which is technologically more challenging (especially the coal gasification technology). In addition, the circulation of gas turbine heat and power cogeneration is used to capture exhaust gas from the gas turbine to make steam for industrial and household use. The thermal efficiency of this kind of circulation can reach 60 percent to 70 percent.

Given China's shortages of oil and gas resources, it is very important to substitute low-quality energy resources for high-quality ones. At present, industrial kilns in China burn approximately 30 million tons of oil a year, and the government intends to cut oil consumption by 10 million tons a year in the coming decade. Therefore, it is imperative to develop technologies to replace oil with coal. The technology of using a water-coal mixture as a substitute for oil, which China has already developed, is gradually maturing. FAW is cooperating with the Shenyang Coal Research Institute to produce high-calorific-value coal gas in place of oil-cracked gas for use in the first domestically and independently designed Luqi furnace for both industrial and household purposes; the technology has proved effective in 2 years of trial operation. Moreover, Luqi furnaces burn lignite as fuel, which is abundant in Jilin Province, thus reducing transportation costs.

We need to actively undertake research on coal gasification technology. Some other countries are developing pressurized moving-bed, fluidized-bed, and gasification-bed coal gasification technology and equipment to put the technology in place to substitute coal for oil and gas in case of either an oil crisis or shortfall of oil or gas resources. Given China's shortages of oil and natural gas resources, it is even more important to develop such technologies. The current coal gasification technology used in China still languishes at the level the rest of the world reached in the 1930s, characterized by low conversion rates and chronic pollution; therefore, it urgently requires improvement.

We must reduce the number of energy conversions, improve energy efficiency, and use industrial steam or gas turbines in place of electric motors wherever practicable. For example, FAW has installed a DA-350-65 turbo air compressor in its power plant with a production of 350 m^3/min of compressed air. The compressor is driven by a back pressure steam turbine, whose exhaust steam is fully used in production. For the year 1987, its electricity consumption (calculated on the basis of steam consumption) per 1000 m^3 of compressed air was 41.08 kW•h in contrast with the average electricity consumption, 106.37 kW•h, of electric compressors, which indicates a remarkable energy conservation effect.

Positive steps are therefore required to develop boilers, fans, water pumps, air compressors, oxygenerators, automobiles, tractors, diesel engines, transformers, electric motors, industrial electric stoves, and welders. These twelve types of products alone account for 33 percent of the country's coal consumption, 90 percent of petroleum consumption, and 54 percent of electricity consumption. Needless to say, improving their efficiency will conserve a considerable quantity of energy.

f. Developing new energy sources

In addition to aggressively developing nuclear energy in areas with a concentration of industrial enterprises but a shortage of energy resources, we need to carry out research into the use of renewable energy sources such as solar and wind power. This approach will not only conserve fossil fuels but will also go a long way toward solving the problem of providing electricity for household use and production in remote areas and on islands.

3. Adopt forceful measures; strengthen the overall management of energy conservation; and use administrative, economic, and legal means to promote energy conservation

a. Energy conservation legislation

China promulgated the Provisional Regulations on Energy Conservation in 1986 and is in the process of enacting the Energy Conservation Law in order to supervise and guide energy conservation actions in accordance with the law. After the oil crisis in 1973, many countries and regions formulated energy conservation laws and regulations, such as the Energy Conservation Act promulgated by the Federal Republic of Germany in 1976; the National Energy Conservation Policy Act enacted by the United States in 1978; the Basic Regulations on Quotas for Fuels, Thermal Energy, and Electricity Consumption in the National Economy promulgated by the Soviet Union in 1979; the Law Concerning the Rational

Use of Energy enacted by Japan in 1979; and the Energy Management Law of Taiwan, China. With the legislation in place, energy conservation work will now be guided by law. At present, we need to make full use of the Provisional Regulations on Energy Conservation and related policies.

b. Strengthening energy conservation agencies

Every region, department, and enterprise needs to assign appropriate agencies and staff to strengthen energy conservation work.

c. Setting energy conservation targets and formulating conservation plans

Every industry and enterprise needs to set energy conservation targets and make plans based on their own record, best practice in their industries at home and abroad, and government requirements. Energy conservation plans should be carefully worked out, especially in the following ten industries: electric power, iron and steel, fertilizer, cement, wall materials, plate glass, aluminum, oil refining, papermaking, and coal, which together account for 41 percent of China's total energy consumption.

d. Using rational allocation of energy resources as an important means of national economic control

Energy resources should be allocated in accordance with China's industrial policy in order to promote industries that require urgent development, restrain industries restricted by the government, and encourage industrial restructuring. At the same time we need to adhere to the principle of assigning high priority to supplying energy resources to certain enterprises and limiting or discontinuing supply to companies consuming excessive quantities of energy.

e. Facilitating energy conservation with such economic levers as pricing, taxation, and credit, and making use of relevant economic policies

Macro-control through the use of economic measures can promote the adjustment of the industrial structure and the product mix, restrict industries that consume large quantities of energy, support those that consume low quantities, reduce energy consumption, and promote the organization of specialized collaboration between industries in such fields as casting, forging, electroplating, and heat treatment—thus eliminating the phenomenon of all-inclusive small enterprises. This problem is relatively serious in China and therefore requires improvement. Furthermore, setting different prices for electricity at peak and off-peak hours can contribute to balancing electricity demand over the course of the day. In Japan this approach has been used to encourage companies to change their schedules. At the Toyota factory the day shift is from 8 AM to 5 PM, and the nightshift from 9 PM to 5 AM, thus avoiding the peak hours of household electricity consumption and making use of the cheaper off-peak electricity after midnight. In China agricultural tractors are in common use for transporting goods, but they are far less efficient than trucks. The policy of low-priced diesel fuel for agricultural use is one of the reasons for this phenomenon.

f. Formulating standards for upgrading energy conservation in enterprises and making this approach a prerequisite for enterprise upgrading

We need to use enterprise upgrading to promote energy conservation. We need to continue to hold energy conservation competitions among different enterprises in the same industry. The policy of substantial incentives and punishment should be adopted, and enterprises and individuals that make outstanding contributions to energy conservation should be rewarded.

g. Organizing research into energy conservation

Every region and department ought to identify its own energy conservation potential and decide what energy conservation measures to adopt. This is a precondition for the formulation of energy conservation plans; therefore, teams of experts should be organized to diagnose problems in major energy-consuming enterprises in order to assist them in energy conservation.

h. Training energy conservation personnel

It is important to equip energy managers and skilled energy conservation personnel in China with the necessary professional expertise. Taking this step will guarantee effective energy conservation. In Japan thermal energy management professionals are trained through a "heat manager" system and must pass a national examination in order to qualify as heat managers. At present this system is on the way to becoming an energy manager system, which places higher demands on energy professionals.

In sum, a great deal remains to be done in the field of energy conservation. We need to mobilize all energy professionals, scientists, and engineers, as well as the general public, to participate in the challenging yet glorious mission of conserving energy. Finally, we need to work hard toward the lofty goal, set by the CPC Central Committee, to quadruple China's 1980 gross industrial and agricultural output value by 2000.

References

[1] International Institute for Applied Systems Analysis. *Energy in a Finite World*. Cambridge, Massachusetts: Ballinger Publishing Company; 1981.
[2] Slesser M. *Energy in the Economy*. New York: St. Martin's Press; 1978.
[3] British Petroleum Ltd. BP *Statistical Review of World Energy* 1985. London; 1986.
[4] National Bureau of Statistics of China. *China Statistical Yearbook*. Chin. ed. Beijing: China Statistics Press; 1984–86.
[5] National Bureau of Statistics of China. *China Statistical Summary*. Chin. ed. Beijing: China Statistics Press; 1986.
[6] Collection of Information on Energy Conservation Research. *Energy Conservation*. Chin. ed. Shenyang; 1984.
[7] United Nations. 1982 *Energy Statistics Yearbook*. New York: United Nations; 1984.

In 1947 Jiang Zemin graduated from the Electrical Engineering Department of Shanghai Jiao Tong University. In March 1989 the university appointed him a professor. After receiving the appointment, he presented an academic report titled "Energy Development Trends and Major Energy Conservation Measures" for teachers and students of the university. After he revised it, it was published in the *Journal of Shanghai Jiao Tong University*, No. 3, 1989.

Nineteen years later, in March 2008, as Shanghai Jiao Tong University was celebrating its 112th anniversary, Jiang Zemin published the academic article "Reflections on Energy Issues in China" in the *Journal of Shanghai Jiao Tong University*, No. 3, 2008. This article expounds on the importance of energy issues and explores the opportunities and challenges facing China's energy development, with a particular focus on the strategic line of thinking with regard to adopting a new, distinctively Chinese energy development path; it also sets forth a corresponding policy approach. The publication of this article is of far-reaching significance for promoting the sustained and sound development of China's energy production and consumption.

After consulting Jiang Zemin and gaining his approval, Shanghai Jiao Tong University asked its research office for university history and Shanghai Jiao Tong University Press to edit and publish a collection of the above two articles, together with the speech Jiang gave on April 9, 2008, to experts attending a forum on "Reflections on Energy Issues in China."

Shanghai Jiao Tong University
August 2008